TURBULENT TRANSFER

IN THE

LOWER ATMOSPHERE

TURBULENT TRANSFER

IN THE

LOWER ATMOSPHERE

BY

C. H. B. PRIESTLEY

THE UNIVERSITY OF CHICAGO PRESS

CHICAGO & LONDON

Library of Congress Catalog Card Number: 59-10427

THE UNIVERSITY OF CHICAGO PRESS, CHICAGO 60637
The University of Chicago Press, Ltd., London W.C. 1

Published 1959. Third Impression 1968. Printed in the United States of America

PREFACE

As a visiting research meteorologist at the University of Chicago in 1957, I gave a series of lectures on the work on turbulent transfer which has been and is being carried out in the Division of Meteorological Physics of the Commonwealth Scientific and Industrial Research Organisation, Australia. It was at the instigation of Professor Horace R. Byers and other members of the faculty of the Department of Meteorology that I later decided to embody the material in book form. When turning from the seminar to the book, one is obliged to take a more comprehensive view, but, in doing so, there is always the danger that enlarged scope and the compounding of other viewpoints may rob the central theme of much of its character of freshness. Accordingly, the temptation to write a book covering the highways and byways of atmospheric turbulence has been resisted and the subject matter restricted to the vertical motions and distributions of the elements and the transports which these achieve. A section on the approach to the geostrophic wind and a short chapter on evaporation have been added to the original material, and the whole has been broadened by including other major contributions of recent years. It is hoped that the results will be both stimulating to active workers in the field and informative to the physicist and meteorologist generally.

For an account of the subject up to and including the contributions in the year 1950 the reader is referred to the excellent textbook on *Micrometeorology* by O. G. Sutton. It is not yet time for this work to be revised, but the time may well be ripe for it to be supplemented, which is the task attempted here. References earlier than 1950 are therefore included only insofar as is necessary to provide the background for the later developments and to produce a whole which is reasonably self-contained; the exclusion of others may mean no more than that they have not been further developed within the compass of this volume.

The work by Lettau and Davidson, *Exploring the Atmosphere's First Mile,* became available to me only after the preparation of this account was well advanced. Whereas it has been tempting to refer to the new experimental results therein reported, especially where these bear on any contentious issue, the dangers of attempting too quick an assimilation of this material suggested omission as a wiser course, but the discussions herein should certainly provide pegs on which to hang analysis of the new data. Nor is the omission a seriously limiting factor, since much of the novelty in what follows stems from measurements of the heat flux which were not paralleled in the Nebraska project and from the analysis of more extreme conditions of instability than were experienced in the Great Plains during the observation periods.

For permission to reproduce diagrams from published papers I am grateful to the Editor, *C.S.I.R.O. Journals* (Figs. 1, 2, 4, 8, and 18), the Director-General of the Meteorological Office and the Controller of H.M. Stationery Office (Fig. 5), the Royal Meteorological Society (Figs. 9, 10, 11, 12, and 16), and the Royal Society (Fig. 20). To my immediate colleagues, whose works are freely quoted herein, I wish to express my further thanks for discussions and helpful criticisms during the preparation of this volume.

TABLE OF CONTENTS

1. INTRODUCTION

A landmark in the development of the study of atmospheric turbulence was the formulation of the concept of the *Austausch* coefficient by W. Schmidt and G. I. Taylor, building on ideas current in aerodynamics and on foundations laid by Boussinesq, Von Karman, Prandtl, and others. An early practical and peculiarly meteorological investigation by Taylor (1915), concerned with the formation of fog when warm air flowed over the Great Banks of Newfoundland, applied the ideas to the vertical diffusion of heat through the lower layers of a thermally stratified atmosphere.

These beginnings served to emphasize that much of the interest of turbulence to the meteorologist lay in its transport properties, notably in the transport of sensible heat, momentum, and water vapor or heat in latent form. The small-scale turbulent flux of these quantities in the vertical generally far exceeds that in the horizontal, particularly in the lower atmosphere, and the vertical flux alone will be considered here, important though the horizontal flux may be in problems on the largest scale, such as the maintenance of the general circulation or the heat and water balance of continents. The values of the vertical fluxes, at a sufficiently low level, represent measures of three of the four main physical exchanges between the body of the atmosphere and its underlying medium, ground or water, the fourth being the radiative exchange. Thus the flux of momentum represents a major and in many ways the most significant external force to the atmosphere, while the other fluxes represent important rates of gain or loss of energy by the atmospheric system as a whole, apart from their direct concern in the practical day-to-day problems of air-mass modification, fog, frost, drought, and so on.

In the generation which followed Schmidt and Taylor the subject of atmospheric turbulence made rapid strides. The development was particularly strong on the theoretical side and in the transposition of wind-tunnel results, though a greater reliance on direct experimental work would have shown that at times the elements of motion have properties markedly different from those of aerodynamic turbulence. During this phase a very large volume of work was concerned with the dispersal of clouds of matter in military and health (pollution) problems. Much attention was also given to the measurement and analysis of the temperature and wind profiles, often as studies in their own right or as indicators of turbulent intensity but not especially directed toward providing new knowledge about surface stress and heat flux and so losing much of their broader meteorological significance. Thus much effort was channeled into specialist aspects or applications, and the meteorologist failed to recognize and fully exploit the new techniques in the wider avenues of his subject. For example, the conversion of the humidity profile into a practical formula for estimating evaporation from natural terrains was largely neglected until 1939 (Thornthwaite and Holzman, 1939), short though the mental distance was from the problem tackled by Taylor twenty-four years earlier.

One may perhaps therefore be excused in pressing the viewpoint herein that, while atmospheric turbulence may be of interest to the aerodynamicist in offering a new scale for his studies, i.e., scope for model experiments in reverse, its importance to the meteorologist rests largely in the knowledge of the transfers which it provides; that the study of profiles is to him primarily a means to this end and that of microstructure a tool for the better understanding of the character of the transfer processes. Since the publication by Sutton (1953) of the first textbook on micrometeorology in the English language, the direct measurement of energy and momentum exchange between earth and atmosphere has been carried through, with associated measurements, bringing new insight into the mechanisms at work and considerably more knowledge about the laws of transfer and the structure of the convective elements; and some considerable reorientation of thinking has followed. Concurrently with this and the development of numerical techniques for the prediction of changes in the atmospheric circulation pattern at large, there is growing recognition of the need for the quantitative application of the interactions between atmosphere and ground to this larger scale. This must inevitably be only approximate and requires to be framed in terms of variables aligned closely with the type of observations which are made, or could be made, at routine observing stations.

There appears much to be gained by attempting a connected account of these aspects now, while a more fully balanced and definitive treatment in textbook fashion awaits the maturity of time. In particular, much clearly remains to be done toward the more effective application in the wider field of synoptic meteorology.

2. THE EDDY FLUX AND ITS MEASUREMENT

Eddy-Flux and Eddy-Transfer Coefficients

Let us consider the vertical transport of some property whose measure per unit mass of air is s, a quantity which will vary from one element of the medium to another. An element of density, ρ, and vertical velocity, w, will then be effecting transport of this property at a rate ρws across the unit area of the horizontal plane in which it is momentarily positioned. Now, using a bar to denote the average over a period of time at a fixed point, the average flux over this period will be $\overline{\rho ws}$, where the bar applies to all elements passing through that point, upward and downward.

Using the prime symbol to denote instantaneous departures from the mean, we may write, following Swinbank (1951),

$$\overline{\rho w s} = \overline{\rho w}\,\bar{s} + \overline{(\rho w)'s'} , \qquad (2.1)$$

a generalization of Reynolds' familiar formulation of the turbulent stress which applies to properties besides momentum and takes full account of fluctuations in density. Thus the flux is composed of a part due to any mean vertical flow of air through the point in question and a part due purely to the eddying motion. At a low height and over a period long enough to embrace a large number of eddy passages, $\overline{\rho w}$ will become small unless there is a very large horizontal convergence or divergence below the point considered, and and this can generally be excluded where the terrain is uniform. The case of sensible heat, where the neglect of the first term on the right is most open to question, is discussed in more detail below. The last term in equation (2.1) represents what is defined as the turbulent or eddy flux, F_s, which may be equated according to convenience to any of the forms

$$F_s = \overline{(\rho w)'s'} = \overline{(\rho w)\,s'} = \overline{(\rho w)'s} . \qquad (2.2)$$

It follows that a direct measurement of eddy flux can be made from instruments sufficiently responsive to record fluctuations in ρw and s over all time scales which contribute significantly to the transfer. Only in recent years has the necessary combination of robustness with the sensitivity required for work within a few meters of the surface been developed along lines described by Swinbank (1951) and elaborated by McIlroy (1955) and by Cramer and Record (1953). Swinbank used an arrangement of hot-wire anemometers to record the wind components, while Cramer and Record used a single one for total wind speed, with a sensitive bivane for the fluctuations in direction. In order to understand the significance of the measurements so obtained, a brief summary of the pre-existing formulations and beliefs is desirable.

The classical formulation regarded an individual element as having "started" its vertical motion, or having last been at rest in the vertical, at a distance l below the reference level at which the flux is considered, and having there had the value of s equal to the average value of the surroundings at that level. Then, if s is conserved in the motion, $s' = -l\, \partial\bar{s}/\partial z$, whence

$$F_s = -\overline{\rho w l}\,\frac{\partial \bar{s}}{\partial z} \approx -\bar{\rho}\,\overline{wl}\,\frac{\partial \bar{s}}{\partial z},$$

the approximation being accurate for all practical purposes, since the percentage variation of ρ from one element to another is altogether negligible as compared with those of w and l. The derivation may be held to apply to the vertical flux, E, of water vapor in terms of the gradient of specific humidity, q; to the flux of momentum or Reynolds' stress, τ, in terms of the gradient of the horizontal wind component, u; and to the sensible heat flux, H, in terms of the gradient of $c_p(T + \Gamma z)$, where T is temperature, c_p is specific heat at constant pressure, and Γ is the adiabatic lapse rate. Moreover, it follows from the derivation that the turbulent transfer coefficient,

$$\overline{wl} = K\ ,$$

is a necessarily positive quantity and one which depends solely on the field of vertical motion, so that K should be the same for the three properties mentioned and for others which satisfy the conditions specified.

The formulation is obviously idealized and is open to a number of objections. One of these is that the "starting" elements need not be an average sample of their environment with respect to the property s but may be subject to bias, particularly in the case of heat, since the buoyancy of the warmer elements constitutes a preferred circumstance. This caused Ertel (1942) and Priestley and Swinbank (1947) to modify the formulation and led to the conclusion that the K for heat should be greater or less than the K for a neutral property under lapse and inversion conditions, respectively. The argument, however, was unable in itself to indicate the magnitude of the difference to be expected, and some other workers (Robinson, 1951; Businger, 1955) have questioned either the principle of the effect or the significance of its magnitude.

A more logical approach is to allow for differentiation between the K's, and write

$$E = -\rho K_w \frac{\partial \bar{q}}{\partial z}, \tag{2.3}$$

$$\tau = \rho K_M \frac{\partial \bar{u}}{\partial z}, \tag{2.4}$$

$$H = -\rho c_p K_H \left(\frac{\partial \bar{T}}{\partial z} + \Gamma\right), \tag{2.5}$$

as purely defining equations for the eddy diffusivities, K_H and K_w, and the eddy viscosity, K_M, without implications as to mechanism. The sign of equation (2.4) is purely conventional in defining the positive direction of stress as that of the downward transfer of momentum. Since the left-hand sides can be measured directly through equation (2.2) simultaneously with determinations of the gradients, the relative and absolute magnitudes of the K's can be determined by experiment.

Comparative Discussion of Measurements

Swinbank (1955) has reported measurements of both τ and H at Edithvale, near Melbourne, on over a hundred occasions, at heights ranging from 1.5 to 29 meters and in a variety of conditions of wind and stability, the selection being biased toward unstable conditions because of the larger magnitude and hence greater meteorological significance of turbulent flux at these times. Cramer and Record (1953) have published similar values from six occasions, making simultaneous measurements at several heights. To these may be added twenty-two measurements of H only, under selected conditions of strong heating and light wind, by Taylor (1956*a*). Measurements of τ and H around 100 meters by Panofsky (1953) and from an aeroplane by Bunker (1955, 1956) will be discussed in later contexts but are less relevant here because of the greater uncertainty in the observed local gradients. Swinbank alone has published measurements of E by the eddy-flux method. Mention might also be made of a small number of measurements of τ and H from a ship at sea (Deacon, 1957*b*), which, while too few at present to justify detailed analysis, are of interest in showing that the additional difficulties associated with such a moving platform can be overcome.

It may be taken as a general principle that the three fluxes are effectively constant with height from the top of the layer, where molecular effects are significant, up to a height of several meters or even, during the main hours of insolation, to 100 meters or so. An exception occurs in the case of H below about 1 meter, for the constancy strictly applies to the sum of eddy and radiative flux, and, since the latter shows significant variations below about 1 meter (Robinson, 1950), the constancy of H applies only from this level upward. These long-established principles can be verified quite simply by calculating that the rates of change of temperature and humidity and the accelerations within the column would otherwise be far greater than are known to occur. It is a main task of this monograph to sort out what is now known about the absolute and relative values and dependence on wind speed and stability of the fluxes and K coefficients, together with the variation with height of the latter. In view of the principle of flux constancy with height and the definitions of K in equations (2.3)–(2.5), the last aspect becomes equivalent to the analysis of the shape of the wind, the temperature, and humidity profiles. Before going into full detail, it is convenient to summarize the broad, order-of-magnitude relationships and the comparative findings of the eddy-flux measurements.

Values of K around 2,500 $cm^2 sec^{-1}$ have been found to be typical at 1.5 meters over grassland in winds of about 4 m/sec and slight to moderate instability, while the variation is broadly in proportion to height and to wind speed. K increases with unstable and decreases with stable stratifications, but not according to any simple general rule. The effects of stability may at times override the other influences, particularly in light winds, when, for example, quite large values of K_H are associated with strong insolation.

From a number of distinct lines of evidence it is clear that, though the thermal stratification $|\partial T/\partial z + \Gamma|$ typically decreases with distance from the surface, its effects on the turbulence and the

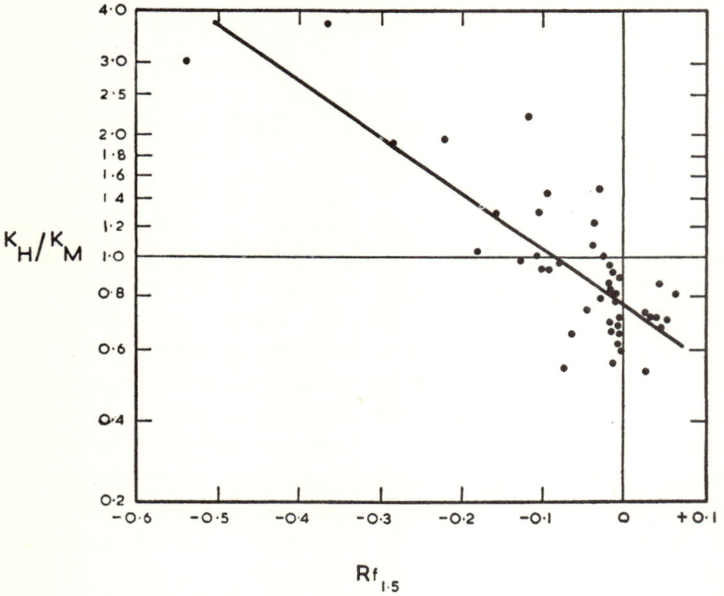

FIG. 1.—Dependence of K_H/K_M on Rf at 1.5 meters

magnitude of the transfer coefficients become increasingly marked with height. The comparison of K_H with K_M from the eddy-flux data at the lowest levels used (1.5 and 1.9 meters) is, then, likely to be the least sensitive for bringing out any dependence on stability; on the other hand, the data there are more plentiful than at other levels and are inherently more accurate, because of sampling errors in the measurement of eddy flux and because the gradients can be measured with greater relative accuracy. Figure 1 shows the individual determinations of K_H/K_M by Swinbank at these levels and also indicates the general trend of the relationship, the scatter being considerable, as would be expected from the nature of the phenomenon. The abscissa chosen to characterize the stability is here the ratio of the rate of consumption of energy by the buoyancy forces to the rate of its production by wind shear. This ratio was the basis of the criterion derived by Richardson in his classical discussion of the rate of growth of turbulence. It is convenient to express it in the form

$$Rf = -\frac{gH}{c_p T \tau (\partial u / \partial z)}$$

and refer to it as the *flux form* of the Richardson number. As may be seen from equations (2.4) and (2.5), it reduces to the more familiar *gradient* Richardson number,

$$Ri = \frac{g}{T} \frac{(\partial T / \partial z) + \Gamma}{(\partial u / \partial z)^2},$$

multiplied by K_H/K_M. The points in Figure 1 give a correlation coefficient of -0.81 between Rf and $\log K_H/K_M$.

In general sympathy with these results, Cramer and Record find values of K_H/K_M which are, if anything, rather less than unity at the lowest level (2.3 meters) under conditions which correspond to small values of $|Rf|$. At their highest level (11.9 meters), K_H/K_M considerably exceeded unity under lapse conditions and was less than 1 on the single inversion occasion.

The magnitude of 0.81 for the correlation is likely to be spuriously high because random errors in the measurement of τ and H would have caused an association between Rf and K_H/K_M in the sense observed. However, Swinbank also found a correlation coefficient of -0.52 between Ri and $\log K_H/K_M$ for the same forty-five pairs of observations, a statistically significant correlation whose magnitude in this case will be spuriously low, since it will be errors in the measured gradients which will now affect both abscissa and ordinate and affect them in a sense opposite to that of the observed correlation. The physical reality of the dependence of K_H/K_M on stability is therefore not in doubt. It is to be stressed that the determinations of τ and H were simultaneous, at the same height, and comparable with regard to instrumental response and sampling errors.

It appears impossible to design a test for the dependence of K_H/K_M on stability which is free from spurious errors, in one sense or the other, of the type discussed above. But it may be seen from equation (2.2) that the eddy-transfer mechanism depends essentially on the correlation between the w and s of the individual elements. So long as we are concerned only to test for differences of mechanism, we may do so quite independently of the measured gradients (and therefore, at the same time, gain relative accuracy at the greater heights) by comparing the correlation coefficients r_{ws}. Furthermore, if any combination of these correlation coefficients is subsequently regressed on Ri, we obtain a test for dependence on stability wherein the two variables under examination have been measured quite independently. Swinbank obtained the following results, over the range $0.1 > Ri_{1.5} > -0.2$:

$$\text{At 1.5 meters} \quad |r_{wT}| - |r_{wu}| = -0.04 - 1.52\,Ri\,,$$

$$\text{At 29 meters} \quad |r_{wT}| - |r_{wu}| = +0.13 - 2.43\,Ri_{1.5}\,.$$

It should be noted that the regression coefficient increases with height and that both values are significant well beyond the $P = 0.01$ level, giving the strongest evidence for a difference in mechanism. The mean values of $|r_{wu}|$ as measured in near-neutral conditions were 0.50 and 0.32 at 1.5 and 29 meters, respectively, the data referring to $0.02 \geq Ri_{1.5} \geq -0.02$ but being weighted toward the superadiabatic side.

The test of K_w against K_M has been made in a somewhat different way. If it be assumed that the two are equal, it follows that the shapes of the wind and humidity profiles will be the same where the fluxes are constant with height and hence that

$$\frac{E}{\tau} = -\frac{q_2 - q_1}{u_2 - u_1} \tag{2.6}$$

in terms of differences between two levels, z_1 and z_2. This has been the basis of a large number of prediction formulas for evaporation, the relation between τ, $u_2 - u_1$, and the heights being supplied from aerodynamic arguments, so that E is obtained from measurements of q_2, q_1, and the wind profile. Since this is the main application of the assumption $K_w = K_M$, it appears to provide a very good background against which to test its validity, and this can be done simply from equation (2.6) without invoking the detail of the aerodynamic theory involved in the individual prediction formulas. In this way Deacon and Swinbank (1956) calculated E from equation (2.6), using measured values of the other quantities, and compared E so obtained with the evaporation directly measured by the eddy-flux method. Here again the values of E and τ were strictly comparable on the basis of sampling and instrumental response. The results are shown in Figure 2, the geometric mean ratio being

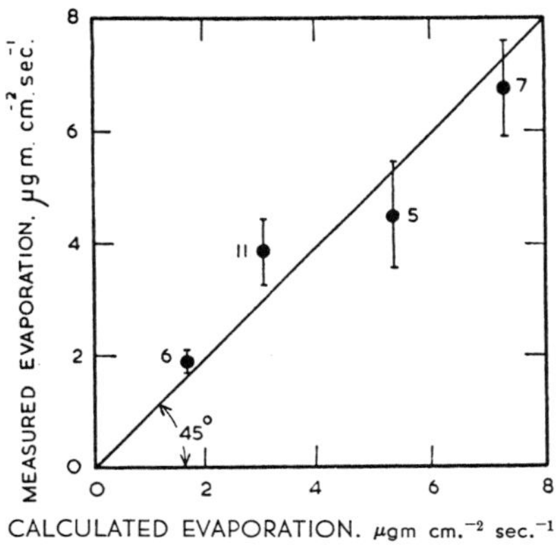

FIG. 2.—Evaporation computed on the assumption that $K_w = K_M$ compared with measured values. The standard errors of the mean values are indicated by the vertical lines.

$$\frac{E\,(\text{observed})}{E\,(\text{calculated})} = 1.04 \pm 0.09\,.$$

The heights used in equation (2.6) were 30 and 0.5 meters, and only the occasions of stronger evaporation were included. A similar procedure applied to data by Rider (1954), who measured the evaporation from small pots and τ by a floating drag plate, gave a ratio 1.12 ± 0.04.

The analysis of K_w against K_M so far described related mainly to a combination of moderate wind with moderate instability. The assumption of equality was shown to provide a satisfactory working rule in what are in many respects the most important conditions. But there are aspects of the problem, such as dewfall or the evaporation in the strong instabilities of the semiarid zones, which generally or always assume importance in conditions well removed from those of the above test. We may calculate, as above, regression equations for the correlation coefficients from Swinbank's data in the range $0.1 > Ri_{1.5} > -0.2$, using all the determinations, irrespective of the magnitude of E (those in brackets in the original paper were doubtful and are not included). We then obtain

$$\text{At 1.5–2 meters} \quad |r_{wT}| - |r_{wq}| = 0.12 - 0.60\, Ri\,,$$
$$|r_{wq}| - |r_{wu}| = -0.15 - 0.76\, Ri\,;$$
$$\text{At 23–29 meters} \quad |r_{wT}| - |r_{wq}| = 0.18 - 1.62\, Ri_{1.5}\,,$$
$$|r_{wq}| - |r_{wu}| = -0.03 - 1.71\, Ri_{1.5}\,.$$

All four regression coefficients have significance levels of P rather better than 0.05, and the increase with height is realistic. On the evidence, the conclusion must be that r_{wq} is intermediate between r_{wT} and r_{wu} in the strength of its dependence on stability. But with measurements of E and r_{wq} fewer in number and in some instances poorer in quality than the other fluxes, there exists a clear need for more measurements and their extension to other climatic types.

Practical Considerations in Measurement

There can be no objections *in principle* to the measurement of flux through equation (2.2); indeed, the method stands alone in this regard and in its independence of the nature of the surface over which the measurements are made. But in all that has been said so far, the question of sampling and systematic errors is a central one. It is relatively easy to meet the requirement that the separation between instruments sensing the different properties should be smaller than the smallest eddies contributing to the flux. It is not so easy, at low heights, to meet the corresponding requirement in the time scale—that the most rapid eddies should be faithfully recorded. Some underestimation of flux due to this cause is to be expected, but it may be kept to the practicable minimum by securing that the sensing-recording system for each property shall have the same effective response time. This need to match the responses requires particular emphasis: if one element of the correlation is definitely limited in some way, "improvement" in the other beyond this limit will throw the two records out of phase, and the final covariance will be less accurate.

The most obvious form of sampling error would arise from the use of too short an averaging period, and, since the size and period of the operative transporting elements increase steadily with height, the sampling error must steadily increase with height unless the sampling period is adjusted in proportion. But the effect is not necessarily conditional on the change of eddy structure with height, as the following argument will show. Whereas, above, it was pointed out that the constancy of flux with height was evidenced by the small value of the acceleration or rate or change of the heat and moisture content of the column, the rates of change referred to were the smooth, maintained values. Within a period of a minute or so the value may be much larger, but random; and, similarly, the flux as measured over such a period would show a random variation with height. In the case of shearing stress this random variation can be quite strong because of the different forms of acceleration which can be and certainly are taken up over short periods. The total amount of momentum or of heat or moisture involved in this random fluctuation process will depend on the depth of layer considered, and, on this account alone, the flux, regarded as a measurement of the mean surface interaction, will decrease in accuracy with height. The much quoted observations of Scrase, who took 1-minute records of τ at 1.5 and 19 meters and found these to be discrepant, provide a good example of this type of sampling error.

Deacon (1955) has gone into this matter in some detail, and a discussion by Swinbank (1955) of the extent to which measured eddy fluxes may be expected to satisfy the energy balance relates to other aspects of the question of site and sampling errors and more generally to the retention or rejection of the mean mass flux term $\overline{\rho w s}$ in equation (2.1). Supposing the flux measurements to be made at height h, where the net downward radiative flux is R, consider the heat balance of a column of unit side along and across the mean direction of the wind. Neglecting diffusion sideways,

$$\int_0^h c_p \frac{\partial}{\partial t}(\overline{\rho T})\, dt = R - G - L - c_p\overline{\rho w T} - \int_0^h \frac{\partial}{\partial x}(c_p\overline{\rho u T})\, dz\,,$$

where u denotes the component in the mean wind direction, G the heat flux into the ground, and L the net heat carried per unit area and time out of the column by water-vapor transfer; the bars throughout will refer to the period over which the flux is measured. Swinbank poses the basic question thus: How far can a set of measurements of the storage term (left-hand side) and of R, G, L, and the *eddy flux* be expected to satisfy the heat balance? Transposing these five terms to the left-hand side, we obtain, after some manipulation,

$$R-G-L-c_p\int_0^h \frac{\partial}{\partial t}(\overline{\rho T})\,dz-c_p\,\overline{(\rho w)'T'}$$
$$=c_p\int_0^h\left[\overline{\rho u}\,\frac{\partial \overline{T}}{\partial x}-\overline{\rho w}\,\frac{\partial \overline{T}}{\partial z}+\frac{\partial}{\partial x}\overline{(\rho u)'T'}\right]dz\,. \qquad (2.7)$$

The first and third terms within the right-hand integral arise from horizontal gradients, and Swinbank indicates that the first will generally be the larger. A temperature gradient of 0.3° C. per 100 meters, in a wind of 5 m sec^{-1}, contributes 2 milliwatts per cm^2 per meter height to the integral. The contribution of the mean motion $\overline{\rho w}$ in equation (2.7) is seen to be associated essentially with the vertical temperature gradient. No means of measuring $\overline{\rho w}$ exist, but the term can be estimated by using the equation of continuity, for, with sufficient accuracy,

$$\overline{\rho w_h}=-\int_0^h \frac{\partial}{\partial x}(\overline{\rho u})\,dz\,;$$

so that this, too, reverts to a horizontal gradient; that is to say, $\overline{\rho w}$ in practice arises as part of a "large eddy." Measurements of $\partial(\overline{\rho u})/\partial x$ are not generally available, but they present no real difficulty; at Edithvale, values of $\partial\overline{u}/\partial x$ in excess of 2×10^{-3} sec^{-1} are seldom exceeded, whence Swinbank estimates a contribution to equation (2.7) generally not greater than 2.0 mw cm^{-2} from the lowest meter. From the magnitudes quoted, compared with typical values of H around 20 mw cm^{-2} during the midday hours, it appears that the heat flux due to eddy motion at 1 meter might differ from that required by the heat balance by up to 20 per cent; but, considering the horizontal gradients likely to exist for some distance downwind of any obstacle which gives shade and shelter, major discrepancies may be expected at sites with poor exposure. Since the integrands are generally of constant sign with height, increasing h will tend to increase the discrepancy.

Sampling and site errors are not, of course, errors in the true sense of the word and in *ad hoc* studies may well become part of the true phenomenon under investigation; but where results of general applicability are required, it becomes desirable to avoid them. Thus the Edithvale site of Swinbank's work was free of obstacles, the responses were matched, the greater attention was given to the lowest level considered practicable (1.5 meters was chosen in this regard), and a 5-minute averaging period was adopted as standard. The adequacy of this was determined by subdividing a few records of longer period and testing for consistency and by subdividing the 5-minute records and repeating the process. This choice has reduced the experimental scatter to the point at which the significant pattern of physical relationships emerges quite clearly at the lower levels; the results from high levels (30 meters), fewer in number and with greater scatter, would not in themselves establish any such clear-cut results but serve to add to the total information in important respects.

The limiting factor in response time for each of the elements ρw, u, T, and q lay in the galvanometers, which were of 2-second free period and critically damped. This was somewhat too slow to record the full contribution to the flux of the more rapid fluctuations at 1.5 meters and so gave rise to a systematic underestimation, which was, in the present view, considerably the most important of all the errors involved. Based on a check of the energy balance, of which all the components were measured throughout the experiments, it was concluded that the measured fluxes were too small by about 30 per cent at this level.

Studies to determine the high frequency loss in greater detail have been carried out by Perepelkina (1957) and by McIlroy (in preparation at the present time). These workers employed oscillographs to record the fluctuations with good fidelity and then applied means of varying duration to smooth the traces progressively and so discover the loss of flux with smoothing. Deacon, in an unpublished manuscript, has analyzed their results and converted them to a form suitable for interpretation in terms of instrumental response time. In Figure 3 is shown the percentage of the full eddy flux, when both component records were smoothed by averages over intervals of length t_m, plotted against $t_m u/z$, this non-dimensional base being chosen because the eddy frequencies may be expected to be proportional to wind speed under given stability conditions and there is evidence to show that the highest significant frequency is inversely proportional to height (Deacon, 1955). Perepelkina's data, for heat flux only, were taken at 1.7 meters height, with $Ri_{1.5} = +0.015$, and are distinguished by a dotted line and square symbols with double tails. McIlroy's measurements were at 1.5 meters, in lapses between $Ri_{1.5} = -0.04$ and -0.07 and single quasi-neutral and inversion occasions with $Ri_{1.5} = -0.007$

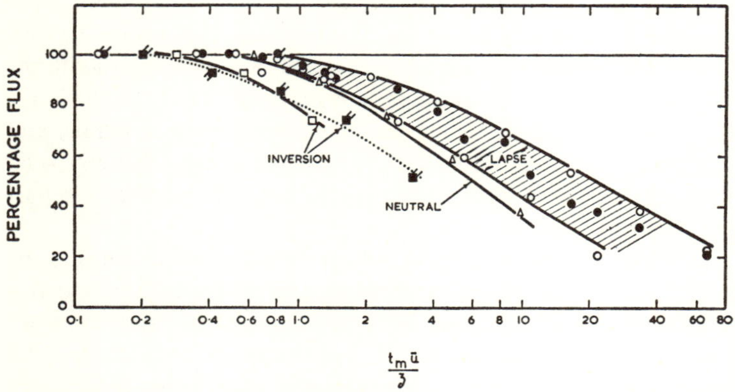

FIG. 3.—Percentage eddy fluxes from records smoothed over interval t_m. Solid symbols for heat flux, open symbols for momentum: ○ ● lapse, △ neutral, □ ■ inversion conditions. For further details see text.

and $+0.04$, respectively. Also included and, as far as can be judged, fitting well with the remainder are values obtained by Panofsky (1953) in lapse conditions at 23 and 91 meters; these are distinguished by round symbols with single tails.

It may readily be shown that a sinusoidal function of amplitude A and period T, when subjected to a moving mean, takes on a reduced amplitude,

$$A f_m = A \left\{ \frac{T}{\pi t_m} \sin\left(\frac{\pi t_m}{T}\right) \right\}.$$

This may be compared with the amplitude reduction factor of an instrument of time constant t_i, which is

$$f_i = \left\{ 1 + \left(\frac{2\pi t_i}{T}\right)^2 \right\}^{-1/2}.$$

Calculation of the two expressions shows that the two reduction factors have a similar type of behavior near $f = 1$ and can be brought closely together throughout the range $1 > f > 0.5$ by setting

$$t_i = 0.4 t_m ,$$

which accordingly provides a satisfactory means of conversion. It is to be remembered that both the moving means and the instrumental smoothing are applied to both factors forming the flux product, so that an amplitude reduction factor of 0.5 corresponds to a 75 per cent loss of flux at the frequency in question. The dissimilarity in the two functions below $f = 0.5$ is therefore of little moment.

A critically damped galvanometer behaves approximately like an instrument of time constant, t_i, equal to one-quarter of the free period, t_g, so that Swinbank's measurements, with $t_g = 2$ seconds, are broadly equivalent to a running average with $t_m = \frac{5}{4}$ seconds. Thus at a height of 1.5 meters and a wind speed of 4 meters sec^{-1}, about the average of the trials, $t_m u/z = 3.3$, which, from Figure 3, implies a loss of flux around 30 per cent in slight lapse conditions, consistent with the deductions from energy balance.

The evidence in Figure 3 for an improvement with increasing instability is consistent with the marked change in scale of eddy structure, discussed in chapter 5.

Figure 3 provides a guide to design criteria for instruments and their heights of operation if the eddy-flux technique is to be used more widely. The need for such wider use is evident, first, to consolidate many aspects of the results discussed in the present work and, second, to tackle certain explicit problems which at present hold little prospect of definitive solution by other means. Among these may be mentioned the stress at the sea surface and the evaporation from the oceans, the evaporation from natural terrains, particularly in the semi-arid regions, and from the taller vegetative covers. To these ends it is desirable, even necessary, to remove the considerable labor at present involved in the evaluation process of equation (2.2) (Taylor and Webb, 1955) so that this is done not only automatically but also concurrently with the fluctuation measurement. This, in turn, requires the response to all fluctuations to be linearized by suitable redesign of sensing elements and circuitry. A linear hygrometer of good accuracy and rapid response has been developed by Taylor (1956*b*). Linearization of ρw and of w can, for practical purposes, be regarded as equivalent, since the relative variations in w greatly exceed those of ρ, and a unidirectional anemometer with this feature (Taylor, 1958*a*) has been incorporated with the hygrometer in an eddy-flux evaporimeter at present under development (Taylor and Dyer, 1958).

3. THE SHEARING STRESS AND THE WIND PROFILE

Shearing Stress in Neutral Conditions

The form of the wind profile and its relation to the shearing stress has by far the largest literature of any aspect of the present subject. It is impossible here to treat exhaustively the many investigations made, and we shall be content to select a few which may best illustrate the current trends of thought and effort. The background of theory and summaries of a large amount of experimental material are to be found in existing textbooks.

The classical experiments of Nikuradse in the laboratory established the existence of two regimes of flow in a turbulent boundary layer. Buoyancy effects were absent, and in this section we shall be exclusively concerned with this condition. *Aerodynamically smooth* flow is said to exist when the profile and τ are related to the kinematic viscosity, ν. At heights above $z = 30\nu/u_*$ the profile is then found by experiment to take the form

$$\frac{\bar{u}}{u_*} = \frac{1}{k} \ln \frac{u_* z}{\nu} + \text{constant},$$

where k is Von Karman's constant, approximately equal to 0.4, and u_* is the *friction velocity*, defined by

$$\sqrt{\frac{\tau}{\rho}} = u_* .$$

Throughout this chapter we shall be dealing with mean profiles, and the bar over the u will be omitted. Below $z = 30\nu/u_*$ the shearing stress is partly turbulent and partly viscous, the former becoming progressively less important as the surface is approached until, below about $5\nu/u_*$, viscosity predominates in what is accordingly termed the *laminar layer or sublayer*.

With increase in wind speed over a given surface or as the surface becomes increasingly rough at constant wind speed, a stage is reached at which the purely viscous stress at the surface is outweighed by the effect of pressure forces associated with eddying wakes from the roughness elements. The flow is then said to be *aerodynamically rough*, and the value of ν ceases to influence the profile, which is found by experiment to depend on a length which is characteristic of the surface roughness. For this regime,

$$\frac{u}{u_*} = \frac{1}{k} \ln \frac{z}{z_0}, \tag{3.1}$$

the ***roughness length***, z_0, being generally an order of magnitude smaller than the actual height of the roughness elements. Aerodynamically rough flow is found to occur when $u_* z_0/\nu > 2.5$. The two profiles are similar in form when ν is replaced by $u_* z_0$, a quantity which Sutton has termed the *macroviscosity*.

Subsequent laboratory investigations have confirmed these laws for a wide variety of sizes, forms, and arrangements of roughness elements. In the boundary layer of the atmosphere a linear relation between u and $\log z$ in adiabatic conditions has been established by a number of workers. At any significant wind speed the numerical magnitudes indicate that the flow will be aerodynamically rough over any vegetated surface, though smooth flow may obtain over snowfields, over mud flats, or over open water in the lighter winds. Sheppard (1947) was the first to show that the value of k conforms to that found in the laboratory, and this matter will be discussed in a wider context below. For the rougher natural surfaces a more precise form of the profile includes the adjustment of z in equation (3.1) by a length z_d, known as the *zero-plane displacement*, which depends on the degree of roughness; but we shall be concerned here with heights much in excess of z_d and will follow the common practice of omitting it from the equation.

Whether the flow is smooth or rough, the shape of the adiabatic profile is uniquely given by

$$\frac{\partial u}{\partial z} = \frac{u_*}{k z}; \tag{3.2}$$

whence, from equation (2.4),

$$K_M = k^2 z^2 \left(\frac{\partial u}{\partial z}\right) = k u_* z. \tag{3.3}$$

It is convenient at times to introduce a drag coefficient, C_D, defined by

$$C_D = \frac{\tau}{\rho u^2} = \left(\frac{u_*}{u}\right)^2 . \qquad (3.4)^{1}$$

Here C_D is closely related to the roughness but, unlike z_0, is not to be regarded as a unique property of the surface, referring as it does to some specified height at which the wind may be supposed known. For reference purposes a table of typical values of z_0 and C_D prepared by Deacon (1953) are shown in Table 1, together with R, the

TABLE 1

ROUGHNESS PARAMETERS OF VARIOUS SURFACES

Type of Surface	R	z_0 (cm.)	C_D at 2 Meters
Smooth mud flats	1.06	0.001	0.0010
Smooth snow on short grass	1.07	0.005	.0012
Desert (Pakistan)	1.085	0.03	.0020
Snow surface, natural prairie	1.10	0.10	.0028
Mown grass:			
1.5 cm.	1.11	0.2	.0034
3.0 cm.	1.14	0.7	.005
4.5 cm. u_2=2 m sec^{-1}	1.185	2.4	.0085
4.5 cm. u_2=6–8 m sec^{-1}	1.15	1.7	.007
Long grass, 60–70 cm. u_2=1.5 m sec^{-1}	1.275	9.0	.016
Long grass, 60–70 cm. u_2=3.5 m sec^{-1}	1.25	6.1	.013
Long grass, 60–70 cm. u_2=6.2 m sec^{-1}	1.21	3.7	0.010

ratio of the wind speed at 2 and 1 meters height. The figures apply to neutral conditions, both R and C_D being dependent on stability, as will be evident in what follows.

For natural surfaces which yield to the wind, such as the longer grasses, the roughness characteristics of Table 1 will vary with the wind speed. It is to be noted that C_D is relatively insensitive to z_0. An equivalent statement is that the drag coefficient can be specified without having to specify too closely the height to which it relates.

Under adiabatic conditions at sea the roughness will depend primarily on the wind speed itself. To the extent that this dependence is complete, the wind profile will contain only a single parameter, the wind speed at one height will fix the whole profile, and it should

1. An earlier convention was to include a factor $\frac{1}{2}$ in the definition, and confusion with values so defined must be avoided.

be possible to plot universal relationships between any pair of variables. The most recent profile measurements at sea (Deacon *et al.*, 1956) have led to the results shown in Figure 4, which are generally in accordance with indications from earlier work of an increase in C_D with wind speeds above 5 meters per second; below that speed, C_D approaches the values appropriate to the aerodynamically smooth profile, which are indicated by the broken line. The sharpness of the increase is still a matter of uncertainty, and more experiment is needed before the best curve can be determined. Corresponding values of z_0 are also shown in the diagram ($k = 0.4$).

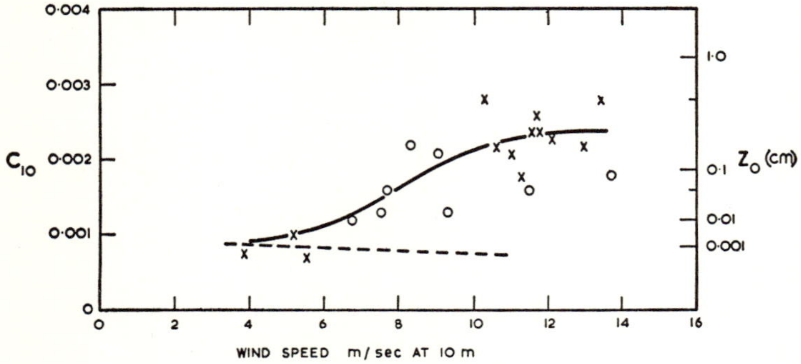

FIG. 4.—Drag coefficient related to wind speed over sea; near-neutral observations, $-0.02 < Ri$ (13 meters : 4 meters) < 0.03.

There is little information on the rate at which the profile will adjust itself to a change in roughness, and the problem is difficult to approach theoretically because the constancy of τ with height breaks down in transitional regions. Over land the situation arises when the wind blows from one surface to another of different roughness. It is often taken as a guide that the profile will be representative up to a height of one-twentieth the distance traversed over the new terrain: this is based on wind-tunnel studies of developing boundary layers and awaits verification on the full scale. Over sea, temporal changes are more significant. It is widely believed that z_0 and τ_0 are governed by the shortest waves or ripples, and, if this is so, the adjustment should be rapid.

THE DIABATIC WIND PROFILE

When the air is being heated or cooled from below, the effects of thermal stratification and buoyancy have also to be considered, so that at least one more basic parameter is required before the state of affairs is specified. The Richardson number has long been used in this regard, and it has been shown formally by Batchelor (1953) that it does, in fact, characterize full dynamical similarity, provided that pressure and density depart by only small fractional amounts from their values in the neutral case and the vertical length scale of the velocity distribution is small compared with $p_0/\rho_0 g$, the "scale height" of the atmosphere expressed in terms of surface pressure and density. These conditions are satisfied in the surface layers of the atmosphere. Since K_H/K_M is one of the quantities which must in turn be governed by dynamic similarity, it is evident that a multiplicative combination such as Rf, or any variant obtained by substitution from equations (2.4) and (2.5), will fulfil the same basic role as Ri. Following Obukhov (1946), Russian workers have used the ratio z/L, where L is a length defined by

$$L = -\frac{u_*^3 \, c_p \rho T}{k g H}. \tag{3.5}$$

The closeness of the relationship between z/L, Ri, and Rf will become apparent in the discussion of results below. Wherever the choice of stability parameter may rest—and this may vary from one to another according to the convenience of the problem in hand—there is no real physical difference in the alternative usages. Whether flux or gradient form is the more truly basic or which is more properly to be regarded as the cause and which as the effect will depend on the situation under study, as will emerge more clearly in the final chapter.

To allow for the effects of stability, one may seek generalization of the conjugate laws (3.2) and (3.3) by writing

$$\frac{\partial u}{\partial z} = \frac{u_*}{k z} f(\xi) \tag{3.6}$$

and

$$\overset{*}{K}_M = \frac{K_M}{u_* z} = \frac{k}{f(\xi)}, \tag{3.7}$$

where ξ is any one of the alternative stability parameters, f is the function which remains to be determined, and $\overset{*}{K}_M$ is a dimensionless form of K_M from which the "neutral" dependence on wind speed and height has been removed. Alternatively, $\overset{*}{K}_M$ can be regarded as a generalized form of the Von Karman "constant."

The most notable feature of the profile shape (Thornthwaite and Kaser, 1943) is that lapse and inversion conditions show opposite curvatures when u is plotted against log z. This characteristic behavior led Laikhtman (1944) and, independently, Deacon (1949) to propose the form

$$\frac{\partial u}{\partial z} = A\, z^{-\beta}, \tag{3.8}$$

where β is less or greater than unity in stable and unstable stratifications, respectively. This constitutes a special hypothesis as to

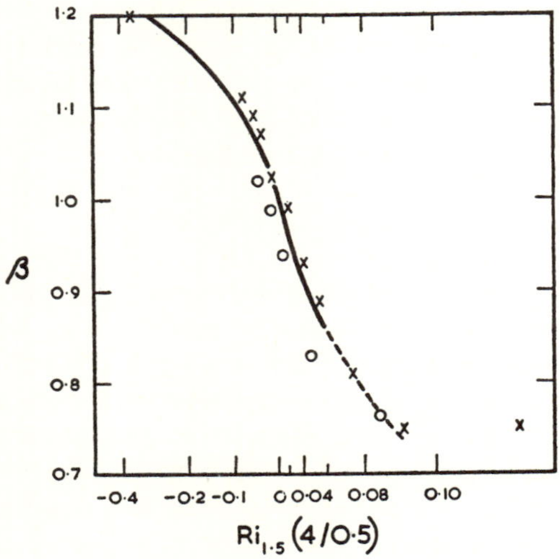

FIG. 5.—Variation of Deacon β with stability

the form in which $f(\xi)$ enters in equation (3.6), it being implicit that both β and the quantity A in equation (3.8) would depend on the parameter ξ. Figure 5 shows the values of β obtained from wind speeds at 0.5, 1, 2, and 4 meters, plotted against a mean Ri for the same layer. Deacon took pains to point out that β was not truly a constant with height, its range of values becoming progressively narrower and converging on unity as the surface is approached. In physical terms this means that at lower heights the effects of stratification become smaller relative to those of wind shear, a conclusion which may be verified by observing that the magnitude of Ri is always small very close to the surface but increases almost linearly with height in both lapse and inversions. Nevertheless, as its wide use attests, the simplicity of form and ready integrability of equation (3.8) make it eminently suitable for summarizing and classifying data and for interpolation.

The integral form is

$$u = \frac{(\tau/\rho)^{1/2}}{k(1-\beta)} \left\{ \left(\frac{z}{z_0} \right)^{1-\beta} - 1 \right\}, \qquad (3.9)$$

the value of the quantity outside the braces being determined by the need to preserve continuity as one progresses into the quasi-neutral conditions close enough to the ground ($\beta \to 1$). Thus, having made use of the condition of near-neutrality at low enough heights, combined with the constancy of τ with height, we arrive at a formula which allows τ to be quite generally estimated from wind-profile observations alone. Where τ is otherwise available, this provides a more stringent test of the Deacon profile than its mere goodness of fit to the wind data. Rider (1954*a*) found satisfactory agreement in all but stable conditions, when the use of equation (3.9) resulted in a serious underestimation of τ. The question has been discussed further by Deacon himself (1957*a*), who attributes the last effect to the absence of fully rough flow. The point is basic not only to this but to other methods of estimating τ from the profile alone, since these all implicitly require the projection of the profile to some low height at which it may be taken that the drag coefficient C_D is dependent solely on the roughness characteristics. Under strongly stable conditions the wind speed at such a level may become small enough for the drag coefficient to depend on the Reynolds number of the flow about the grass blades, and fully rough flow will not then exist. The generalization required to cover this contingency must recognize the Reynolds number as an additional basic parameter of the flow, even though it will play a passive role over a wide range of conditions.

Although not avoiding this last difficulty, the most fundamental framework yet proposed for the analysis and unification of the wind profile is that of Monin and Obukhov (1954). The roughness length, essentially a constant of integration of equation (3.6), can be eliminated by integrating between two assigned levels. The generalized or universal wind profile is then most readily expressed as a relation between differences of u between two heights and the ratio of these heights. The essence of the new approach was to employ the similarity principle, identifying one of these heights with the scale length $|L|$, or $|L|/2$, defined as above. Monin and Obukhov observed profiles from 0.5 meter (in some cases 0.2 meter) to 15 meters and set out to express the results as a single relationship between $u(z) - u(|L|/2)$ and z/L. They do not, however, measure L but a quantity B, to which L is assumed to be uniquely related. B is defined as

$$B = \frac{g}{T_0} \frac{T(z_1) - T(z_3)}{u^2(z_2)},$$

where the heights are chosen as $z_1 = 50$ cm, $z_2 = 1$ meter, $z_3 = 2$ meters; B is accordingly an inverse length and plays the role of a Richardson number divided by height.

For small values of z/L, equation (3.2) may be generalized at once by writing

$$\frac{\partial u}{\partial z} = \frac{u_*}{kz}\left(1 + \alpha \frac{z}{L}\right), \tag{3.10}$$

where α may be expected to be a universal constant. This assumption was first introduced by Halstead (1943) but without developing the similarity considerations. Integration of equation (3.10) gives

$$u(z) = \frac{u_*}{k}\left(\ln \frac{z}{z_0} + \frac{\alpha z}{L}\right). \tag{3.11}$$

The data were gained from four expeditions, which require separate analyses in view of differences in roughness length, z_0. The profiles were first grouped according to the value of B and averaged and were then fitted by least squares to equation (3.11) so as to determine the best values of the parameters α/L and u_*/k for each mean profile and the best single value of z_0 for the site. It appears, though it is not entirely clear, that only data from the lowest 4 meters were used in the fitting process, but even then a considerable number of points were used for which z/L was not small enough to rely on the validity of equations (3.10) and (3.11).

If the temperature profile is assumed to follow the same law as equation (3.11), it may be shown that B is given as $1/\alpha$ times a function of α/L which is known in terms of z_1, z_2, and z_3, and α was next obtained by regressing this function on the measured values of B. They obtained $\alpha \approx 0.6$, but this may be substantially in error (see, e.g., the discussion below). Having assigned the values of all the parameters in this way, it was possible to plot each point of every profile in a diagram of

$$\frac{k}{u_*}\left[u(z) - u\left(\frac{|L|}{2}\right)\right]$$

against z/L, with the result shown in Figure 6.

While the exploitation of the similarity principle represents a step forward in the analysis of wind data, certain points must be borne in mind when appraising the implications of Figure 6. First, it is B, not L, which is in reality the basic element of the diagram, the transformation from one to the other having been based on the assumption that the profiles of temperature and wind and the associated eddy mechanisms are identical, thereby setting aside important issues. Second, the asymptotic relations indicated for large $z/|L|$ are not consistent with equation (3.11), which was used in deriving them, so that there may be some systematic distortion at either side of the diagram; for example, the wind shear does not disappear at such small negative values of z/L as visual inspection of the figure would suggest, as may be verified by reference to other data or even to the raw data on which this diagram is based. The choice of a linear scale for z/L instead of the more usual logarithmic one may here be at fault, and it certainly is at fault in so condensing the region close to the z-axis that the physical relevance therein is lost. The power to mislead of a visual appeal to profile shape, when plotted on a linear height scale, is already notorious, for it becomes almost impossible to distinguish between one profile having, say, a twenty-fold change and another having a thirty-fold change in slope between two fixed levels. This means (see eqs. [2.3]–[2.5]) that a 50 per cent error in K or in the ratio of the two K's may be completely hidden, and it illustrates a basic weakness in the attempted use of profile shape as a discriminant in any sensitive test. The consequence here is that Figure 6 fails to provide any information on one of the most important questions, that of the limits of stability or instability within which the neutral laws may be taken, for practical purposes, to apply. However, the question may be answered by reconciling the approach with that of Deacon, noting that

$$\frac{z}{L} = \frac{k}{\overset{*}{K}_M} \cdot \frac{K_H}{K_M} Ri = \frac{k}{\overset{*}{K}_M} Rf .$$

So long as conditions remain effectively neutral, $k/\overset{*}{K}_M$ and K_H/K_M both remain equal to or close to unity, and z/L is equivalent to Ri. Figure 5 shows that the effects of stratification are significant well within the range $+0.05 > Ri > -0.1$, and it may therefore be

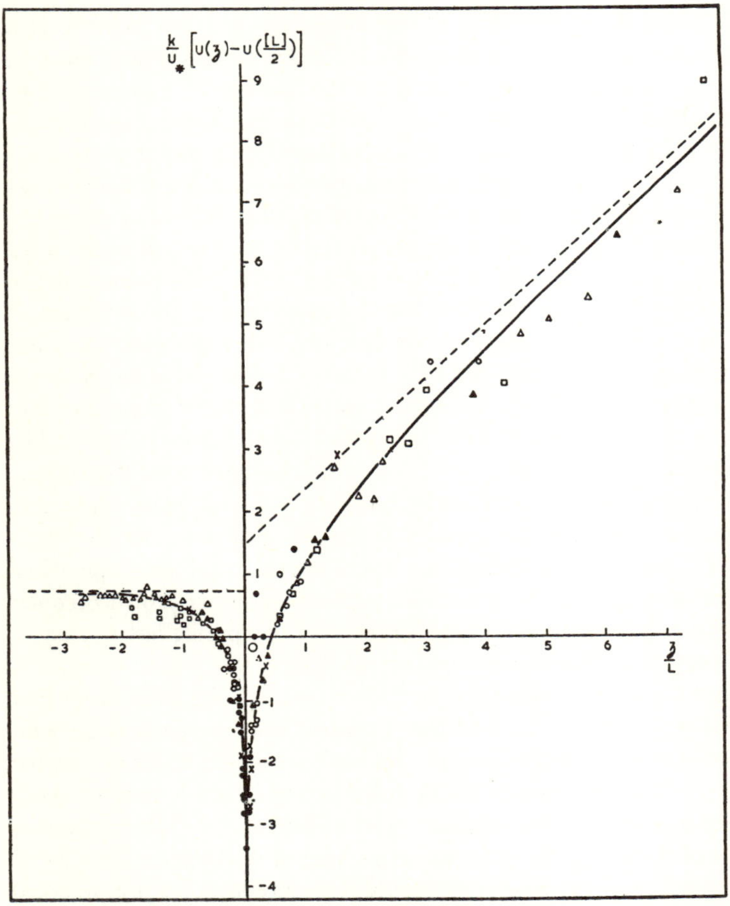

FIG. 6.—Universal wind profile of Monin and Obukhov

concluded that this holds also within the same range of values of z/L.

The manner of processing the Monin-Obukhov data has to some extent restricted the freedom for reappraisal of the results, but some partial verification of the last conclusion is possible. The impression was given in the original of Figure 6 that effects of thermal stratification become significant only for $|z/L| > 0.5$, because log-

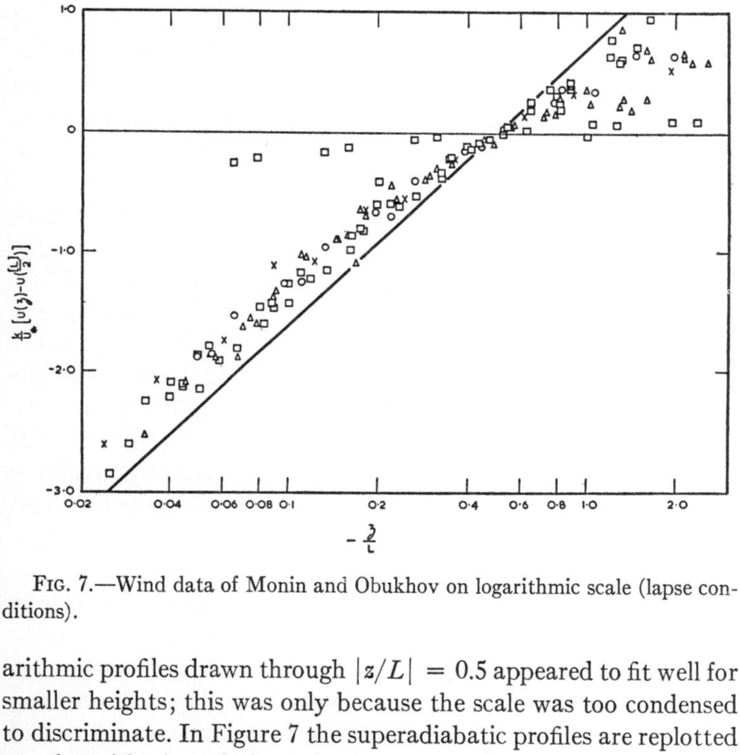

FIG. 7.—Wind data of Monin and Obukhov on logarithmic scale (lapse conditions).

arithmic profiles drawn through $|z/L| = 0.5$ appeared to fit well for smaller heights; this was only because the scale was too condensed to discriminate. In Figure 7 the superadiabatic profiles are replotted on a logarithmic scale from the data set out in tables in the original paper. Different symbols relate to different years of observation, there being several mean profiles in each year according to the value of B, with six levels in each profile; of some twenty-five profiles, two (*open squares*) are readily identified as anomalous, possibly due to a misprint in the data. The logarithmic profile (3.2) is drawn as a full straight line, and the systematic departure from it both above and well below $|z/L| = 0.5$ is evident by inspection.

Shearing Stress in Diabatic Conditions

The Deacon and the Monin and Obukhov frameworks for the analysis of the diabatic wind profile discussed above have been selected from others, the first as being the simplest and currently the most widely used, the second as the most fundamental. When it comes to the practical problem of estimating the stress on a given occasion, they have much in common, since each, in its most reliable application, is based essentially on measurement of both wind and temperature gradient and calculation of the Richardson number therefrom. Deacon's method can, at its face value, be freed from the need to measure any thermal quantity, since β and the other quantities in equation (3.9) can be determined from the wind profile alone, but this is not to be recommended as a practical procedure. The physical reality and magnitude of the departure from neutrality depend on the rate of change with height of $z(\partial u/\partial z)$, and with present techniques this cannot be measured sufficiently accurately and reliably to provide a useful practical index for the individual occasion. It appears likely that the use of the shape of the profile as a stability index will long remain a matter of principle rather than practice, and the utility of the wind profile toward the stress will continue to be in the measurements of the gradient which it provides. Even when measurements of temperature profile are added to those of wind, it has been seen that knowledge of the stress has been built up only by inference and in a way that leaves considerable room for uncertainty.

A more direct and satisfying type of analysis can be carried through when the experimental data include both τ and the profiles. Rider (1954*a*) has employed the technique first developed by Sheppard (1947), wherein surface stress is determined by measuring the force on a floating plate made to simulate the natural ground surface; the wind profile was measured simultaneously, and values for $\overset{*}{K}_M$ were calculated from equation (3.7) over a range of thermal stratifications. Deacon (1955) has treated existing eddy-flux measurements in the same way. The plot of $\overset{*}{K}_M$ against Ri is shown in Figure 8. The aim of such a diagram is to portray the effect of stability on the transfer coefficient after the first-order dependence on height and wind strength has been removed. Here open circles represent points obtained by Cramer and Record, filled-in circles those of Swinbank at 1.5 meters, the standard error being indicated by short-line segments in instances where it is possible to average six or so individual determinations in a narrow range of *Ri*. The broken line is the root-mean-square curve of Rider's results.

As verification of the diagram in extrapolation, Deacon has subsequently analyzed eleven stress measurements by Panofsky (1953) at levels around 100 meters and found $\overset{*}{K} = 1.00 \pm 0.12$. *Ri* cannot

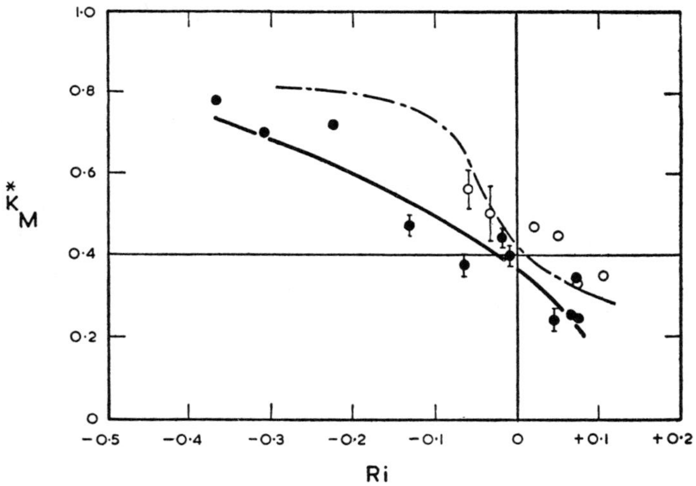

FIG. 8.—Variation of $\overset{*}{K}_M$ with Richardson number

be identified accurately but was well beyond any values plotted in Figure 8; the average *Rf*, where it could be calculated, was −0.9.

Holzman (1943) has proposed the relationship

$$\overset{*}{K}_M = k\,(1 - \sigma_2 Ri)^{1/2}\,, \tag{3.12}$$

which, though quite empirical, provides the simplest algebraic form that might be expected to hold over the range in question, though it clearly must fail at large enough stabilities. Conversely, a relation suggested earlier by Rossby and Montgomery for stable conditions,

$$\overset{*}{K}_M = k\,(1 + \sigma_1 Ri)^{-1/2}\,, \tag{3.13}$$

must fail in the more unstable regimes, and the Holzman form is therefore preferred as a framework for discussing the present data.[2] For the best fit to Swinbank's data, Deacon finds $k = 0.37$, $\sigma_2 = 8.2$, and the full curve in Figure 8 represents Holzman's equation drawn accordingly. The value of k within 10 per cent or so of that obtained in the wind tunnel constitutes an interesting cross-check between field and laboratory techniques. Deacon has discussed quite fully the results obtained from all attempts to measure or infer this important quantity in the atmosphere and, bearing in mind some consistent underestimation in Swinbank's measurements as discussed in the previous chapter, concludes that the best value is close to 0.41.

With regard to dependence on stability, the difference between the sets of data is not so large as Figure 8 suggests because averages were taken in one case, and root mean squares in the other, from original data which were subject to large individual scatter. Regression of the data on Holzman's expression yields

$\sigma_2 = 8.2 \pm 0.7$ (eddy stress, Swinbank, gradients at 1.5 meters) ,

$\sigma_2 = 7.2 \pm 1.3$ (drag plate, Rider, gradients at 1.5 meters) ,

$\sigma_2 = 8.5 \pm 0.7$ (drag plate, Rider, gradients at 0.75 meters) .

Thus it may be concluded that the over-all variation is broadly established by experiment within the range $-0.3 < Ri < +0.1$, to which the above regressions refer. But the data are not yet sufficiently accurate and plentiful to define the exact slope of the curve in any smaller part of the range. For example, on the basis of the solid points, which appear the more accurate, a line of zero slope would fit equally well from $Ri = 0$ to about -0.05; a reanalysis of Rider's original data on the basis of average rather than root-mean-square $\overset{*}{K}_M$ also gives a much flatter curve just to the left of $Ri = 0$. The question is relevant to the parallel discussion of heat-flux measurements in chapter 4.

A point deserving repetition is that, apart from the use or lack of use of measured fluxes, there is rather little difference in essentials between the various approaches described in this chapter and between them and some others excluded here only for reasons of brev-

2. But see chap. 7.

ity, and it is only to be expected that a recent critical comparison (Elliott, 1957) indicates close similarity in the nature of the findings. Besides their common merits, they have a common failing, in that the treatment of K_M and τ in their dependence on stability is entirely empirical. There is no accepted body of theory to provide a framework for the analysis of results or a guide as to the form of relations to be looked for. At one extreme some guidance might be provided by the classical theorem of Richardson (1920) that beyond $Rf = +1$ the turbulence will decay, so that turbulent transfer in the sense considered here will no longer take place. But it would be unwise to assume that the relationships corresponding thereto would be approached asymptotically from the turbulent side. The tendency for turbulence to increase or decrease on opposite sides of the limit argues that nature will strive to avoid this value, and any change to or from the turbulent state would be abrupt rather than gradual.

The present basis for the treatment of K_W and E is no less empirical, but, essentially because of the closeness of the link between temperature gradient and stability, the treatment of K_H and H can be based more securely on physical principles. It is thus possible to discuss the transfer of heat much more fully in the chapters which follow.

The Approach to the Geostrophic Wind

Whereas we have so far been concerned with the magnitude of the momentum flux at the surface and in the layers very close to it, where the flux is sensibly constant, the problem of approach to the geostrophic wind involves deeper layers and is governed by the manner in which the momentum is absorbed therein as a function of height. In unaccelerated horizontal motion the velocity vector $\boldsymbol{V}$ is given by the equation

$$\lambda \rho \boldsymbol{k} \times (\boldsymbol{V} - \boldsymbol{J}) = \frac{\partial}{\partial z}\left(\rho K_M \frac{\partial \boldsymbol{V}}{\partial z}\right), \qquad (3.14)$$

where λ is the Coriolis parameter, $\boldsymbol{J}$ the geostrophic wind, and $\boldsymbol{k}$ a unit vertical vector. From this point, two courses are open. Assuming some model for K_M as a function of z, we may solve equation (3.14) for $\boldsymbol{V}$, also as a function of z, thus providing complete theoretical relationships between the wind and the isobars, which may then be checked against what is known from observation. Conversely, we may adopt a position of ignorance about the transfer mechanism and appeal to the height integral of the above equation, or

$$\tau_h - \tau_0 = \lambda \int_0^h \rho\,(u_J - u)\, d\,z\,, \qquad (3.15)$$

where τ is the vertical flux of any component of momentum and u_J and u are the components of J and V at right angles to this direction. Observations of geostrophic departure may then be used to provide information about τ.

The first line of work has been well summarized by Sutton (1953), and the chief advance since that time has lain in the increasing recognition of the complications of the problem. The classical study with constant K resulted in the familiar Ekman-Taylor spiral, and subsequent more elaborate models for K have adduced some numerical changes in the relations therein without affecting the general pattern. Nearly all the treatments take J to be a constant, i.e., they ignore the influence of the thermal wind. The most realistic of the models is considered to be that of Rossby and Montgomery (1935), who assume a surface-layer law of the form of equation (3.3) to hold up to some height h and thereafter a steady decrease of K to a value of $A = \rho K = 50$ gm cm^{-1} sec^{-1} taken to represent the residual turbulence of the free atmosphere. A recent analysis by Ellison (1956), in which equation (3.3) is taken to hold throughout, may be a step away from realism on this account, but it does serve as a reminder that certain properties of the boundary-layer solution may be carried over into the new context. The drag coefficient, i.e. $(u_*/J)^2$, now in terms of J, the magnitude of the geostrophic wind, is shown to be relatively insensitive to roughness length, z_0; and again over sea, where there should be a strong interdependence between u_* and z_0, we may suppose the whole profile in neutral conditions to be determined approximately in terms of any single one of the parameters u_*, z_0, J.

With a number of K models having led to more or less complicated analytical solutions of the wind profile through equation (3.14), there has been a tendency to ask for definitive measurements capable of distinguishing between them. A more appropriate approach, however, is to inquire whether any typical state of affairs can be expected to exist and what generalizations can be considered worthwhile, in view of the range of variation which the thermal wind regime and the daily incidence of convection and other symptoms of thermal stratification can impose. The discussion of observations of westerlies over the sea by Sheppard *et al.* (1952) is particularly relevant here, in showing, among other things, that the vector variation of wind from 200 to 500 meters is accountable as thermal wind and that the surface winds in this regime blow virtually parallel to the isobars. None of the existing solutions of equation (3.14) could possibly have predicted these features, because none of them has recognized the possible dominance of the thermal wind. With regard to convection, in parts of southern United States in summer the diurnal variation of wind in the surface layers is very considerable, reaching a maximum at around 2,500 feet (Buajitti and Blackadar, 1957), with a total range of some 90 to 40 per cent of the mean speed, respectively, along and across the direction of the mean vector. A similar effect is known to occur in Australia and is probably typical of the continental regime in low and middle latitudes. Its study obviously requires the introduction of a time dependence for K_M, and Buajitti and Blackadar conclude that such variations can occur only when both the average value and the amplitude of K_M decrease rapidly with height above about 250–300 meters. Thus, while the type of generalization commonly introduced in studies of the true boundary layer of 100–200 meters is acceptable because the effects of space and time variations are not strongly interactive, the danger of extending such generalizations to deeper layers has been well brought out by the authors of the last two papers quoted.

In the higher latitudes and less continental climate of Quickborn, near Hamburg, data by Frankenberger (1952) suggest that the effects of thermal stratification on the wind shear are only slight at levels above 70 meters. On clear summer days the wind speed at this level has quite a pronounced diurnal variation from 75 per cent of its mean value in the morning, when the inversion dissolves, to 120 per cent around midnight. Generally, however, the ratio of geostrophic to 70-meter wind speeds lies between 1.5 and 2.0, whereas that of geostrophic to 10-meter speed rises from 2.5 in neutral to about 7 in stable conditions. Knowledge of the latter ratio is particularly relevant in bridging the gap between the profiles of the layer discussed in this section and the more purely micrometeorological data with which the rest of the chapter is concerned; since wind speed at around 10 meters is a routine synoptic observation, the data for further analysis of the ratio already exist in abundance.

To obtain an explicit formula for K_M, we may multiply equation (3.14) vectorially by $\boldsymbol{V}$, and it follows immediately that

$$\rho K_M = A = \frac{1}{V^2 (\partial \alpha / \partial z)} \int_0^z \lambda \rho (V^2 - VJ \cos \alpha)\, dz\,, \quad (3.16)$$

where V and J are the magnitudes of $\boldsymbol{V}$ and $\boldsymbol{J}$ and α is the angle between them. It has been pointed out by Lettau (1950) that the results so obtained may be very sensitive to the azimuth adopted for $\boldsymbol{J}$ and, one may add, equally sensitive to its magnitude in a different set of circumstances; in short, the geostrophic wind as taken from the weather map is too inaccurate for the evaluation of equation (3.16). In certain conditions the difficulty may be overcome and $\boldsymbol{J}$ obtained from the profile itself. Lettau describes a method, essentially one of trial and error, which applies when $\boldsymbol{J}$ can be treated as a constant and provided that the profile displays the main features of the classical, spiral form. Orienting the x-axis in a direction tentatively identified with that of $\boldsymbol{J}$, one may identify heights z_1 and z_2 where components v and u, respectively, first reach maximum values and where it is assumed that τ_y and τ_x, respectively, first vanish. From the two component equations of equation (3.15) it then follows that

$$\tau_{0,x} = \rho\lambda \int_0^{z_2} v\, dz\,,$$

$$J = \frac{1}{z_1} \left\{ \int_0^{z_1} u\, dz + \tan \alpha_0 \int_0^{z_2} v\, dz \right\},$$

since $\tau_{0,y} = \tau_{0,x} \tan \alpha_0$. These can be evaluated from the profile data. From the calculation one can finally recover values of A from $\tau_x/(\partial u/\partial z)$ and $\tau_y/(\partial v/\partial z)$, and these will be equal only when a correct choice of α_0 has been made.

Applying these methods to the careful and detailed wind measurements made by Mildner at Leipzig, Lettau found α_0 to be 26° and from this was able to determine the values of A set out in Table

2. The general form of variation is not unlike that assumed in the original model of Rossby and Montgomery.

With reference to the determination of τ_0 from equation (3.15), calculations from the Leipzig profile by Lettau gave $\tau_0 = 5.3$ dynes cm^{-2} for a geostrophic wind speed of 17.5 m sec^{-1}, corresponding to a drag coefficient of 1.4×10^{-3} in terms of J. In terms of surface wind speed over the ocean in the tradewind belt, Sheppard and Omar (1952) found C_D centered around 1.0–1.5 $\times 10^{-3}$ at three places. These last are lower than values obtained from other work and, moreover, show no variation with wind speed in the range from 3 to 12 m sec^{-1}. A fourth station gave values from 9 to 2 $\times$

TABLE 2

LETTAU'S ANALYSIS OF THE LEIPZIG WIND PROFILE

	HEIGHT (METERS)							
	50	100	200	300	400	500	700	900
Wind component* (m sec^{-1})	9.15	10.45	12.60	14.30	15.62	16.83	18.23	18.68
Wind component† (m sec^{-1})	4.35	4.64	4.95	4.90	4.60	4.00	2.73	1.31
A (gm cm^{-1} sec^{-1})	132	163	179	179	166	150	118	70

* Parallel to J. † Perpendicular to J.

10^{-3}, decreasing with increasing wind speed. The mean of 466 double-theodolite balloon ascents made over 27 days at Anegada in the Virgin Islands gave $C_{10m} = 1.24 \times 10^{-3}$ for a wind of 5 m sec^{-1} (Charnock, Francis, and Sheppard, 1956).

One of the difficulties of the method is that, before a component of $\boldsymbol{\tau}_0$ can be calculated from the geostrophic departures, the value of that component of $\boldsymbol{\tau}$ at some other level, h, must be assigned. Where westerly thermals overlie surface easterlies, there is generally a well-defined maximum wind within a few thousand feet of the surface, and the assumption may be made that the appropriate component of τ is zero at this level. There is little either to verify or to disprove this assumption at the present time, but it certainly becomes less secure when applied, as in the studies quoted above, to the mean of a group of profiles within which the level of maximum varies widely. In the baroclinic westerlies over sea there is no well-marked maximum, and the assigning of τ_h becomes even more dubious. Another difficulty, not unrelated in the latter case, is that of determining J and its variation with height with sufficient accuracy to remove the possibility that a substantial contribution to the integral in equation (3.15) may come from small geostrophic departures maintained through deep layers above the level at which they are commonly assumed to vanish. Over land, other things being equal, both τ_0 and the cross-isobar flow at low levels are larger, and the method may be expected to give percentually more accurate answers than over sea; the error, which is represented by the momentum flux into the deeper atmosphere, through the levels of a maximum component or those otherwise conventionally taken to mark the top of the so-called friction layer, is a smaller fraction of τ_0. This fraction, however, plays a significant part in the momentum balance of the general circulation, and further discussion of errors in the present context must lead into this wider subject.

The method of estimating τ_0 from equation (3.15) is seen to admit both systematic and casual errors and to be particularly vulnerable in the context of the over-all momentum balance. On the other hand, it provides an approximate value, likely to be useful in more restricted applications, and requires only the type of data that is obtainable without specialist techniques. Moreover, it is certain that we cannot afford to neglect this approach to the problem, since it provides an estimate of the drag of a *terrain* rather than that of a *site* to which the finer-scale methods refer. Over land, at least, the distinction is an important one.

4. HEAT CONVECTION AND THE TEMPERATURE PROFILE

Free and Forced Convection

Convection of heat or of any other property denotes the transfer from one place to another by the motion of the medium, and, as stated in the Introduction, we are here concerned with the vertical transfer only. *Free convection* is said to occur when the motion, or that part of the motion which carries the heat, is itself set up by buoyancy; *forced convection* when the motion is otherwise imposed.

In fully forced convection, by definition, buoyancy does not affect the motion or the heat-transfer coefficient, which in turbulent motion is the quantity K_H already defined, and the dynamically generated turbulence is taken to be the only agent in transferring heat down the temperature gradient. The motion itself must therefore be that of the neutral regime defined in chapter 3, from which we may obtain the following boundary-layer relations:

$$K_H = K_M = k^2 z^2 \left(\frac{\partial u}{\partial z}\right), \tag{4.1}$$

$$H = -\rho c_p K_H \left(\frac{\partial T}{\partial z} + \Gamma\right) = -\rho c_p k^2 z^2 \left(\frac{\partial u}{\partial z}\right)\left(\frac{\partial T}{\partial z} + \Gamma\right), \tag{4.2}$$

$$\left(\frac{\partial T}{\partial z} + \Gamma\right) \propto z^{-1}, \tag{4.3}$$

the last deriving from the neutral profile, equation (3.2), and the constancy of H with height. For reference purposes, with $\partial u/\partial z = 1\ \text{sec}^{-1}$ at 1 meter height, $K = 1{,}600\ \text{cm}^2\ \text{sec}^{-1}$; and with $\partial T/\partial z + \Gamma = -10^{-2}\,^\circ$ C. cm^{-1}, $H = 19.2$ mw cm^{-2} = 0.276 cal cm^{-1} min^{-1} ($k = 0.4$, $c_p = 0.24$, $\rho = 1.2 \times 10^{-3}$).

The corresponding laws of free convection in the atmosphere were derived from local similarity considerations by Priestley (1954*a*) as a particular case of the more general physical problem of the free convection from heated objects. From the definition it follows that the laws cannot contain wind speed or wind shear explicitly, but, despite this simplification, the general problem is not determinate because there remain seven independent variables on which the heat flux may depend. As always in dimensional analysis, the identification and grouping of the independent variables is not unique, but, provided that it is done in a physically relevant way, the choice is not critical. A suitable set consists of a characteristic length, temperature difference (or gradient), specific heat, density, gravity multiplied by coefficient of thermal expansion (here g/T), kinematic viscosity, and thermal conductivity. Now in the atmospheric case the main buoyant elements are so large that ranges of smaller-scale motions intervene between them and the molecular scale. The mixing between the buoyant elements and the environment becomes itself a turbulent type of process, and molecular transfer coefficients will no longer appear in the relationships. Then, writing

$$H \propto \rho^{\alpha} c_p^{\beta} \left(\frac{g}{T}\right)^{\gamma} z^{\delta} \left|\frac{\partial T}{\partial z} + \Gamma\right|^{\epsilon},$$

we may equate dimensions in terms of fundamental units of length, mass, time, temperature, and heat, and obtain

$$H = h\,\rho\,c_p \left(\frac{g}{T}\right)^{1/2} z^2 \left|\frac{\partial T}{\partial z} + \Gamma\right|^{3/2}, \qquad (4.4)$$

where h is a constant whose value remains to be determined. The assumption has been made that $\partial T/\partial z + \Gamma$ is the relevant gradient for a compressible fluid and that all temperature profiles in free convection are similar, which follows as a special case of Batchelor's (1953) theorem. In the words of Monin and Obukhov (1954), who derived the law independently by arguments of a similar nature, a regime of purely thermal turbulence is self-patterning, there being no scale length provided by the independent parameters of the situation.

Inverting equation (4.4), we obtain the temperature profile in the form

$$\frac{\partial T}{\partial z} + \Gamma = -\left(\frac{H}{h\,\rho\,c_p}\right)^{2/3} \left(\frac{T}{g}\right)^{1/3} z^{-4/3} = -C z^{-4/3}, \qquad (4.5)$$

where C is, for practical purposes, constant with height and may be used as an index of magnitude of the temperature gradient. We may also write

$$K_H = h\, z^2 \left(\frac{g}{T}\left|\frac{\partial T}{\partial z}+\Gamma\right|\right)^{1/2} = h\left(\frac{gC}{T}\right)^{1/2} z^{4/3}\,. \qquad (4.6)$$

Laws (4.4)–(4.6) can, of course, be obtained from more mechanistic assumptions. They were first derived by Prandtl (1932) from the classical mixing-length approach, adding the assumption that individual rising elements experience a balance between buoyancy force and a resistance proportional to the square of the upward velocity.

As with the wind profile of forced convection and its associated relationships, laws (4.4)–(4.6) cannot be extrapolated right down to the surface because turbulence is not in control there. It was pointed out in chapter 2 that radiation exchange exerts an important influence up to about 1 meter, and the relevance of the present discussion is confined to the region above that level. For reference purposes in equations (4.4)–(4.6), with $T = 295°$ A and $\partial T/\partial z + \Gamma = -10^{-2}$ ° C. cm^{-1} at 1 meter and other constants as before, $C = 4.64$ cgs ° C. units, $K/h = 1{,}820$ cm^2 sec^{-1} at 1 meter, $H/h = 21.9$ mw cm^{-2} = 0.315 cal cm^{-2} min^{-1}.

Heat-Flux Measurements

Equations (4.2) and (4.4) form the background of analysis when measurements of H are available with simultaneous observations of wind and temperature gradients. It is convenient for this purpose to make H dimensionless by writing

$$\overset{*}{H} = \frac{H}{\rho\, c_p\, (g/T)^{1/2} \,|\, (\partial T/\partial z) + \Gamma \,|^{3/2} z^2}\,,$$

so that the flux-gradient relationships for forced and free convection appear in simple algebraic form, respectively, as

$$\overset{*}{H} = k^2\,|Ri|^{-1/2} \qquad (4.7)$$

and

$$\overset{*}{H} = h\,. \qquad (4.8)$$

It must be noted that $\overset{*}{H}$ will be a constant with height within the free-convection, but not within the forced-convection, regime.

From Swinbank's measurements under lapse conditions the values of $\overset{*}{H}$ were calculated and plotted against $|Ri|$ on logarithmic scales, as shown in Figure 9 (Priestley, 1955). It was possible for this purpose to make use of eighty-four such pairs of measurements, for, although when H was measured at the greater heights the value of Ri at the same heights was not available, Ri was referred to 1.5

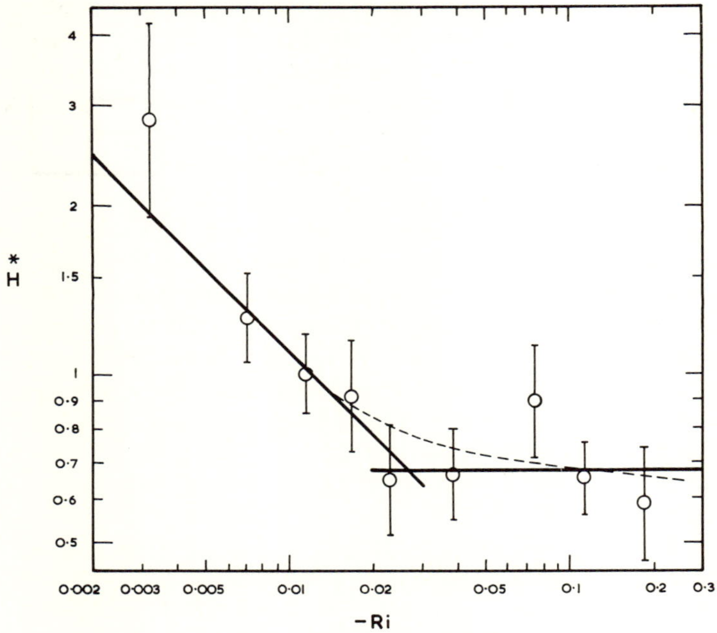

FIG. 9.—Dimensionless heat flux, $\overset{*}{H}$, versus Ri from Swinbank's measurements. Mean values for groups are shown by circles, and standard (log) errors by the vertical line segments.

meters on each occasion and the measured value of H could be regarded as applying there also.

To the extent that equations (4.7) and (4.8) define extremes between or within which the true regime should lie, the downslope of the relationship in Figure 9 must lie within the bounds $\frac{1}{2}$ to zero and should decrease progressively from left to right. The outcome of the analysis was the suggestion that it changed rather rapidly from one extreme to the other, the lines in the diagram being the straight lines of best fit, respectively, to all points for which $|Ri| < 0.03$ (accepting the exclusion of a line of downslope greater than $\frac{1}{2}$) and for which $|Ri| > 0.02$. The uncertainty was clearly too large to allow any definite conclusion to be drawn, as illustrated by the broken smooth curve which fits the data virtually as well. However, the concept that the heat-flux law can be even approximately represented by two distinct regimes rather than by a gradual range of intermediate ones is a simplification whose practical advantages require little elaboration. Furthermore, the suggestion that an effectively free-convection regime may extend from $Ri = -0.02$ or $Ri = -0.03$ is one which in itself has important implications, both practical and theoretical, some of which can be subjected to independent test.

One of the most direct and stringent tests of the hypothesis implicit in Figure 9 is its use in extrapolating values of the heat flux under more extreme conditions than those of the original data, i.e., the horizontal line not only should remain but should be unaltered in position within experimental error. Taylor (1956*a*) made twenty-two further measurements of H under conditions with values of $Ri_{1.5}$ between -0.175 and -0.9, and his combined plot of all data is reproduced in Figure 10. The straight line sloping upward represents the regression on all data for which $|Ri| > 0.02$. The degree of slope (0.05 ± 0.05) was not statistically significant, and the broken line shows the horizontal line of best fit. For the mean of all such observations Taylor found

$$\overset{*}{H} = h = 0.69 \pm 0.05 \; ,$$

while for those involving H measured only at heights of 1.5–4 meters, which, for reasons described in chapter 2, are the more accurate, he found

$$\overset{*}{H} = h = 0.79 \pm 0.04 \; .$$

That a total of seventy-six measurements over the range $-0.02 > Ri > -0.9$ shows no trend with Ri and thus no correlation with wind shear must be taken as substantial evidence for the existence of an effectively free-convection regime.

No other measurements of H of comparable scope and quality have been made, but Kasanski and Monin, in a private communication (to be published), have reported findings from the heat balance in broad agreement but with somewhat larger values of h. This would accord with the considerations of chapter 2; evaluation of $t_m u/z$ and use of Figure 3 indicate that, on the whole, the free-convection measurements of H should be some 10 per cent too small. We shall accordingly adopt

$$h = 0.9 \text{ *}$$

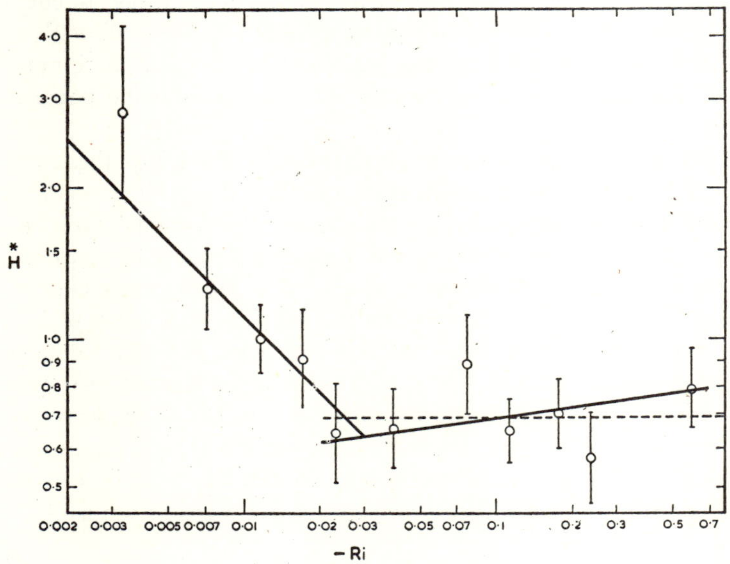

FIG. 10.—$\overset{*}{H}$ versus Ri from both Swinbank's and Taylor's measurements

as the best value. With T, ρ, etc., given the same values as above in equation (4.4), this corresponds, in the convenient form used by Kasanski and Monin, to

$$H = 0.17\,|\Delta T|^{3/2}\,,^{*} \tag{4.9}$$

where H is in cal cm^{-2} min^{-1} and ΔT is the difference in temperature (° C.) between 2 and $\frac{1}{2}$ meter.

Accepting the implication of the existence of two regimes, it is useful to define the Richardson number, Ri_c, which approximately locates the transition, without suggesting anything in the nature of

* See note page 52.

a discontinuity there. From equations (4.7) and (4.8) the intersection of the two straight lines is given by

$$-Ri_c = \frac{k^4}{h^2}, \qquad (4.10)$$

which, with $k = 0.4$ and $h = 0.9$, is equal to 0.03. The lines as drawn in Figure 10, where both are subject to any underestimation of flux which may be present in the observations, intersect at about this value. Here again it would be inappropriate to insist on an accurate value, and the original suggestion of Priestley (1955) that transition is effected between -0.02 and -0.05 still appears to represent about the right degree of tolerance.

The outstanding feature of the results still calling for an explanation in theory is the observed numerical value of h or, equivalently, that of Ri_c. This cannot be provided by dimensional and similarity arguments but requires the placing of the theory on a mechanistic basis. Whereas any model of the free-convective motion which assumes a unique relationship between heat flux and local gradient must lead to an expression of the type (4.4), the values of h therein will depend on the details of the model, and h may well be regarded as a measure of the efficiency in heat transfer of the model in question. Buoyant elements which mix rapidly with their environment will, as they move upward, retain less buoyancy and acquire less momentum than those which mix more slowly and so, other things being equal, will be less efficient. Thus continuing elements or *plumes* may be expected to be more efficient than bursts or *bubbles* and line elements more efficient than point elements, since in each case there is one less dimension in which the mixing is free to operate. It is not certain that nature will incline to select the most efficient model consistent with the restrictive conditions of the problem (e.g., the condition for continuity of mass), but it seems reasonable to expect that it will do so. But it would be premature to pursue these conjectures beyond the preliminary stage until we have described more fully what is known from observation of the actual structure of the convective motion, and this topic will be taken up in chapter 5.

From the practical viewpoint it is to be noted that diagrams like Figures 9 and 10 provide a straightforward method of arriving at estimates of H from measurements of the gradient. Determination of Ri fixes both the regime and the value of $\overset{*}{H}$, whence H can be calculated. As a guide to the incidence of the transition, Table 3 has been prepared from occasions of clear skies on summer days at Edithvale, in winds (at a height of 2 meters) of different strength. Thus on either an average or a majority basis we may regard the transition at 1.5 meters as occurring typically at wind speeds around $4\frac{1}{2}$ m sec^{-1}, or $5\frac{1}{2}$ m sec^{-1} if the wind speed is referred to normal anemometer level. These values relate to latitude 38° S. and a typical H of about 20 mw cm^{-2}.

TABLE 3

VALUES OF $Ri_{1.5}$ ON CLEAR SUMMER DAYS AT EDITHVALE

	U_2 (M SEC^{-1})					
	0–2	2–3	3–4	4–5	5–6	>6
Per cent of $\|Ri\|_{1.5}<0.03$	0	7	22	52	63	100
Per cent of $\|Ri\|_{1.5}>0.05$	100	83	78	41	0	0
Average $\|Ri\|_{1.5}$	0.37	0.18	0.10	0.05	0.02	0.015

THE SUPERADIABATIC TEMPERATURE PROFILE

Independent tests of the existence of the two regimes with a well-marked transition between them may be made from analyses of temperature profiles and their conformity, or otherwise, to laws (4.3) and (4.5). Priestley (1954*a*) was able to show that on fine summer days around noon the published data on temperature profiles from Ismailia (Egypt), Porton (England), Manor (Texas), and Poona (India) were broadly consistent with equation (4.5) from 1 or 2 to 20 meters or so. However, in the previous discussion of wind-profile shape there was occasion to notice the difficulty in obtaining fine discrimination, particularly from published data not always appropriately grouped, and the same applies to the temperature profile.

Figure 11 shows the results of an analysis by Webb (1958) of 111 profiles at Edithvale, each profile being of 5 or 10 minutes' duration under cloudless daytime summer conditions. The diagram has $Ri_{1.5}$

as abscissa and, for ordinate, employs a shape characteristic of the temperature profile defined as

$$\frac{1}{150 \ln \frac{8}{2}} \cdot \frac{[T + \Gamma z]_{2m}^{8m}}{(\partial T/\partial z + \Gamma)_{1.5m}}$$

where the bracketed term denotes the difference between values at the superscript and subscript levels. The characteristic has the value unity for the forced-convection law $\partial T/\partial z + \Gamma \propto z^{-1}$ and 0.73 for the free-convection law $\partial T/\partial z + \Gamma \propto z^{-4/3}$, as indicated by the

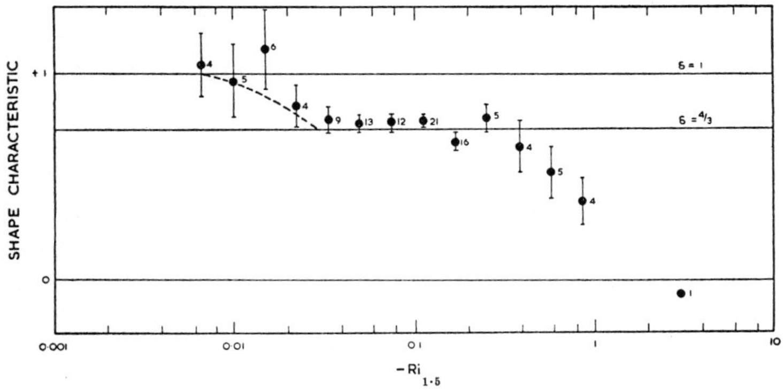

FIG. 11.—Shape characteristic of the temperature profile between 8 and 1.5 meters.

two horizontal lines. The interval between them is crossed by a segment of slanting, broken line which denotes the values that the shape characteristic would take if the z^{-1} law held for both wind and temperature gradient up to the level at which $-Ri = 0.03$, and the $z^{-4/3}$ law for temperature gradient beyond. It is seen that the temperature profile supports the evidence for an extended regime of free convection beyond $|Ri| = 0.03$ and is consistent with a transition to forced convection around that value, though the data are too few and scattered to indicate its sharpness. A gradual transition would bridge the gap between the two horizontal lines by a curve of gentler slope than the one drawn. There is an interesting suggestion of a discrepancy beyond about $Ri_{1.5} = -0.4$, and we shall return to this question later.

Another consequence of the typical occurrence of free convection at strong enough lapse rates would be the existence of a practical upper limit to the lapse rate at any assigned height above ground. This is not a new idea in principle, for such an upper limit has been calculated in the well-known Rayleigh-Jeffreys criterion for the breakdown of a layer into regular cellular motions of the Benard type. However, at the levels with which we are here concerned, regular cells are seldom, if ever, observed, and the criterion so obtained appears irrelevant not only because its assumption of constant lapse rate is so far from reality but also because other, more irregular, convective motions are likely to be set up first and then prove capable of carrying away the heat as fast as it is supplied. Since there is an upper limit to the incoming solar energy, there will also be an upper limit to H and hence, through equation (4.5), to C. Measurements have been made by Vehrencamp (1953) under a high sun on a dry lake bed in California which must represent a very close approach to the extreme condition. Assuming evaporation to be zero and measuring the other components in the heat balance, Vehrencamp inferred a heat-flux rate of 0.7 cal cm^{-2} min^{-1}, which corresponds to $C = 8.5$ cgs ° C. units when $h = 0.9$. That this does give a representative upper limit to the lapse rate is supported by reference to sites where strong lapse rates are to be expected and where continuous records have been taken. At Ismailia (Flower, 1937) the average of extreme lapses between 16.2 and 1 meters for the months of April to September inclusive was 3.1° C., with only minor differences between the individual months. Between 39 and 4 feet at Suffield in semiarid Alberta (Tait, 1949), half-hourly values of temperature differences in summer and autumn showed quite a well-defined maximum between 5.0 and 5.1° F. Here $C = 8.5$ corresponds to 3.4° C. in the first case and 5.1° F. in the second.

Well-marked passages from one convection regime to another are common features of experimental work in the laboratory, and there is no reason to doubt that they will have counterparts in the atmosphere, provided that the conditions are satisfied which correspond to good laboratory control: i.e., steady heating (this may exclude intermittent sunshine) and no intrusion of advective effects from irregularities of site. Nature in practice will exhibit all degrees of intermittency and irregularity, and the type of study described here can hope only to depict the fundamental modes of behavior and demonstrate that these do exist in the right circumstances. From independent lines of evidence we have noted a substantial body of support for the existence of a free-convection regime, subject to laws (4.4)–(4.6) and beginning at about $Ri = -0.03$. If this is true, the dual problem of superadiabatic temperature profile and convective heat flux is solved in principle, and in practice there would remain only the question of the right degree of smoothing for a given site and occasion—a question which, like many in meteorology, might have to remain a matter of judgment rather than of reference to explicit rules. It is not difficult to understand why a large mass of previously published data on temperature profiles has failed to indicate any special preference for a $\frac{4}{3}$-power law; to nature's smoothing has been added the smoothing of analysis, averaging over considerable height and time intervals, and classifying with respect to time of day without reference to Ri or other physically relevant index.

The operation of a free-convection regime at appreciable wind speeds is not, of course, to be construed as implying the absence of mechanically generated turbulence but merely that the latter becomes insignificant in heat transfer. The viewpoint may be taken that the general level of H is determined by the gross parameters of the situation (insolation, wetness and general nature of ground, etc.) operating through the principle of energy balance, and that near the ground the mechanically generated turbulence can carry the heat away as rapidly as the situation requires. But its intensity declines with increasing height, and at some level, which will be lower the lighter the wind and the stronger the heat flux, it becomes inadequate for the task, and free convection must take over. The relative rapidity with which this occurs may be mainly a reflection of the self-regulating property of the free-convection process.

We return now to the one discordant feature of the picture so far obtained, namely, the suggestion in Figure 11 of a departure from the $\frac{4}{3}$-power law at sufficiently large $-Ri$. To test for reality, it is necessary to extend the data to more extreme conditions or, equivalently, to greater heights on the same occasions, and Webb repeated his analysis, as in Figure 12, with the ordinate now appropriate to a higher layer, i.e.,

$$\frac{1}{150 \ln \frac{30}{8}} \frac{[T+\Gamma z]_{8}^{30}}{(\partial T/\partial z+\Gamma)_{1.5}},$$

the lines having a significance corresponding to that in Figure 11. What is now revealed is a region of much reduced or even vanishing gradient of potential temperature, beginning at a height of about 30 meters when $Ri_{1.5} = -0.1$ and descending to around 8 meters when $Ri_{1.5} = -0.4$. Bearing in mind the normal, approximately linear,

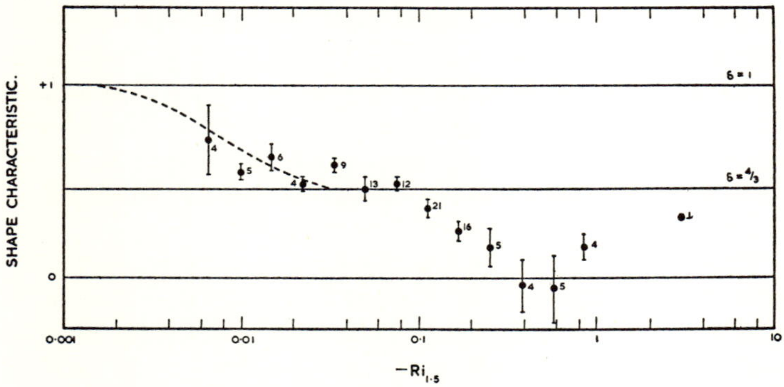

FIG. 12.—Shape characteristic of the temperature profile from 30 to 8 meters.

increase of Ri with height, we may speak broadly of a second transition at about $Ri = -2$, which again appears to be a rather sharp one, to a state of affairs at which the superadiabatic condition almost or actually vanishes. The extreme negative values of $Ri_{1.5}$ are distinguished from others which are large but not extreme by lightness of wind and wind shear rather than by the magnitude of $\partial T/\partial z + \Gamma$: in other words, on a given occasion of strong insolation it will be primarily the wind speed which governs the heights of the transition. In sympathy with this finding, it may be noted that the maximum lapse rates at Ismailia, referred to above, were observed in moderate, not in light, winds.

This feature of Webb's temperature profiles cannot be verified in the same detail from the large amount of observational material collected by other workers, owing to the choice of different parameters for the grouping of data, but Webb has at least been able to show that they are not inconsistent. Measurements on high towers show that, with strong heating and light winds, the lapse rate averaged through the higher layers falls to a temporary minimum in the hours around noon, when that in the lowest layers is a maximum. This is found at Karachi (Mal *et al.*, 1942), at Rye in southeast England (Best *et al.*, 1952), and by Flower at Ismailia. At Quickborn, near Hamburg, zero gradient or even slight inversions are found above 30 meters on clear summer days with light winds (Frankenberger, 1953). A deep layer of vanishing temperature gradient has also been observed in the laboratory in convection between parallel plates (Thomas and Townsend, 1957), but, although the laboratory experiments were in a still chamber, the region in the atmosphere is not one of vanishing wind shear. Probably "vanishing" is too strong a word for either context, but, whereas the temperature gradient becomes too small to measure over depths of 10–20 meters, the wind shear in general does not.

The assumption underlying the derivation of equations (4.4)–(4.6) was that the local heat flux was related directly to the local gradient. Since the breakdown occurs at levels where H must sensibly retain its full value, it is evident that this assumption is no longer satisfied. There must, in other words, be strong interaction between remote parts of the flow. Certainly, from the wider meteorological viewpoint, such interaction is only to be expected. Here we may consider the deeper layers of atmosphere which are more strongly mixed, or at least have much smaller vertical gradients, as comprising a sink of large heat capacity and the bulk difference in temperature between the sink and the surface source as influenced by processes on a larger scale than that appropriate to the boundary layer. This aspect will be taken up in chapter 8; it is sufficient for the present to note that equations such as (4.4) and (4.5) must be specifically boundary-layer equations. The exact nature of the transition at the upper limit of their validity remains to be identified. Current work, not yet published, offers one hope for explanation, in that the transition represents a change in the nature of the mixing between the buoyant element and its surroundings from a regime in which this is governed by the turbulent properties of the whole field to one where it is more dependent on the element's own characteristics. In the latter case the impetus received from below will influence the mixing rate at higher levels, and the same local similarity will not apply.

Whatever may be the exact nature of the more remote interactions, they must be such as to adjust themselves so that a quasi-constancy of H is preserved. The nature of the mixing between an element and its environment will figure prominently in the discussions of chapter 6. On it depend the ability of the individual buoyant element to penetrate and the question of the linkage between cloud and ground, leading naturally to other problems of convection through the deeper layers.

NOTE ADDED 1965

A subsequent program of heat flux measurements with improved instruments and a reassessment of the systematic errors in the earliest results confirm the regime obeying the laws (4.4) to (4.6) but give a significantly higher value of $h = 1.32 \pm 0.06$: correspondingly, 0.17 in equation (4.9) would become 0.25. See DYER, A. J. 1965. "The Flux-Gradient Relation for Turbulent Heat Transfer in the Lower Atmosphere," *Quart. Jour. Roy. Meteorol. Soc.*, **91**:151–57.

5. THE SPECTRUM OF TURBULENCE AND THE STRUCTURE OF FREE CONVECTION

GENERAL CONSIDERATIONS

The turbulence near the ground varies from one situation to another not only in intensity but in character, and previous chapters have shown how both these factors may enter into the flux-gradient relationship. For the purpose of specifying the character of the turbulence, the mutually related methods of correlogram and spectral analysis have now become virtually standardized. Often, however, quite simple averaging or mere visual inspection of records will serve to illustrate a typical pattern of structure or behavior, and a selection of such examples will serve as an introduction to the more sophisticated mathematical analysis.

Figure 13 shows two segments of records obtained at Edithvale of simultaneous fluctuations of temperature and horizontal wind speed at a height of 1.5 meters, taken on a clear day and the following clear night without change in wind speed. The remarkably high degree of correlation between the two elements of each pair of records must be due to the dominance of some connecting link—clearly, the vertical motion. This component will be assigned a central place in the statistical analyses which follow.

A second important feature of the eddy structure can be discerned in Figure 13, but is much more clearly brought out in Figure 14, where 3-minute records at 29 meters of i, the inclination of the wind to the horizontal, are shown on the same scale for another day-and-night pair of occasions. The wind decreased somewhat from 4 to 3 m sec^{-1}, but the main physical change lay in the reversal of $Ri_{1.5}$ from -0.12 to $+0.04$. The inversion condition displays a small and rather rapid irregular fluctuation about a constant (zero) inclination. In the lapse condition this type of turbulence is present with little change but now appears imposed as an embroidery on a larger and much slower irregular fluctuation. It is natural to assign the feature which is common to the two records to the turbulence-producing influence common to the two occasions—i.e., wind shear—and to assign the feature which is confined to the upper record to buoyancy—the influence which operated on that occasion and not on the other. As will appear, the difference is characteristic. The two types of motion become clearly distinguishable when smoke

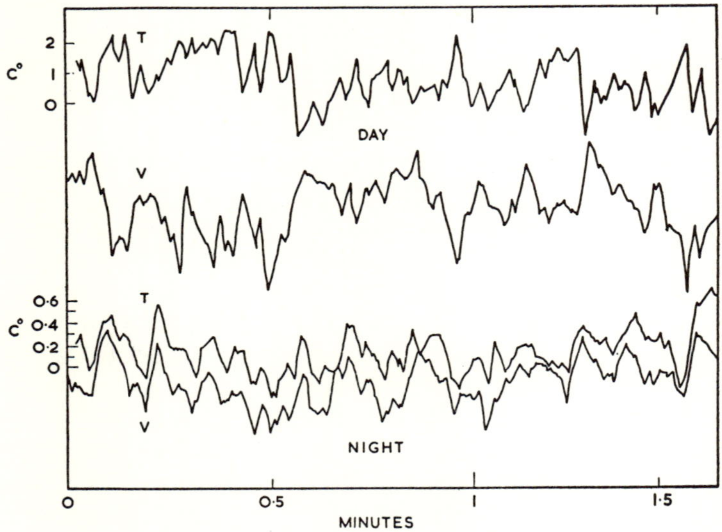

FIG. 13.—Simultaneous records of T and V at 1.5 meters, wind speed 2.7–2.9 m sec^{-1}.

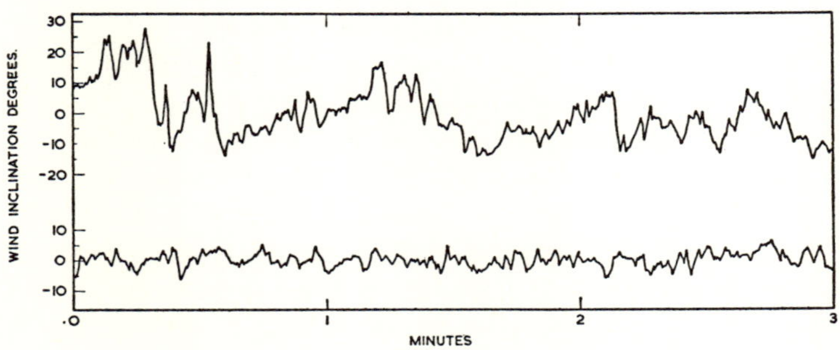

FIG. 14.—Three-minute records of wind inclination at 29 meters. *Top*, u = 4 m sec^{-1} in lapse; *bottom*, u = 3 m sec^{-1} in inversion.

is released from a continuous source, one type serving mainly to diffuse the smoke while the other produces the effect known as "looping," wherein considerable lengths of the cloud are carried bodily upward or downward. The vertical distance traveled by these loops shows further that the slower fluctuations of Figure 13 not only are of somewhat larger amplitude locally but are much more deeply penetrative than the more rapid ones.

The third example is taken from the work of Panofsky (1953*a*). Records over an hour in length of w, T, and horizontal wind speed, V, were taken at heights of 75 and 300 feet on the tower at Brook-

TABLE 4

RELATIVE CONTRIBUTIONS OF DIFFERENT EDDY SIZES TO ENERGY AND FLUX

		Small Eddies	Medium Eddies	Large Eddies
Kinetic energy	75 ft......	60	29	7
	300 ft......	35	35	19
Heat flux	75 ft.......	46	26	27
	300 ft......	13	26	61
Stress	75 ft.......	55	48	− 3
	300 ft......	15	64	21

haven National Laboratory on an occasion of slight lapse and moderately strong wind (4–7 m sec^{-1}). Ten-second averages of all the variables were obtained, and the w values were further smoothed by taking successively 70-second and 490-second moving averages. The procedure does not achieve a clear-cut resolution in terms of period but, broadly speaking, serves to separate the w fluctuations into those occurring in periods between 20 and 140 seconds, between 140 and 980, and longer than 980 seconds. These smoothed w records were then each multiplied by T, V, and by themselves, to provide the relative contributions from the three periods to the heat flux, stress, and vertical kinetic energy, respectively, with the results shown in Table 4. It is characteristic of a larger body of data that the small eddies contribute less to the heat and momentum fluxes than to the energy, and the largest eddies more to the heat flux than to the others. The marked increase in eddy size with height is also characteristic.

The mathematical analysis aims to sort out the attributes of the turbulence of the type discussed above and apportion them to discrete points or ranges along the scale of period or frequency. When resorting to this device, however, it must be borne in mind that there is a gain in formality and precision but not always in physical relevance, and at times more is to be learned from a visual study of an original record than from its mathematical transformations. From continuous records at a fixed point of two variables, say u and w, with respect to time, t, the Eulerian time-correlation function,

$$r_{uw}(\sigma) = \frac{\overline{u'(t)\, w'(t+\sigma)}}{\sqrt{\overline{u'^2(t)}\ \overline{w'^2(t)}}}, \qquad (5.1)$$

is first formed. We define $F(n)$ such that

$$\{\overline{u'^2}\ \overline{w'^2}\}^{1/2} F_{uw}(n)\, dn$$

represents the contribution to the covariance $\overline{u'w'}$ (with $\sigma = 0$) contained in the frequency band between n and $(n + dn)$. Then $F(n)$ is given by

$$F(n) = 2\int_{-\infty}^{\infty} r(\sigma) \cos 2\pi n \sigma\, d\sigma, \qquad (5.2)$$

and is known as the *co-spectrum* of u and w. The *energy spectrum*, defining the contribution between n and $(n + dn)$ to $\overline{u'^2}$ or $\overline{w'^2}$, may be obtained as a special case, and equation (5.2) can then be written as follows:

$$F(n) = 4\int_{0}^{\infty} r(\sigma) \cos 2\pi n \sigma\, d\sigma,$$

the inverse relationship being

$$r(\sigma) = \int_{0}^{\infty} F(n) \cos 2\pi n \sigma\, dn. \qquad (5.3)$$

The integral over all frequencies is given by

$$\int_{0}^{\infty} F(n)\, dn = r_{uw}(0) \qquad \text{for the co-spectrum},$$

$$= 1 \qquad \text{for the energy spectrum}.$$

The merit of this form of analysis over the ordinary Fourier series is that it deals in a relevant way with fluctuations which recur with only approximately the same form and period and with randomized phase shifts.

In consequence of the absence of any upper limit to eddy sizes or periods in the atmosphere, the total energies and total fluxes will depend on the period over which the average is taken, and so too will the functions $r(\sigma)$ and $F(n)$, as may be readily seen from equation (5.1). This difficulty can be met only with the rather loose convention that the period must be chosen appropriately to the time and space scales of the problem under consideration. For problems occurring within the time scale of the diurnal variation, periods ranging from around 5 minutes at 1 meter to 1 hour at 100 meters are generally considered most suitable.

Different ranges of the spectra may relate to quite different energy and flux processes. Turbulent energy in the system is created from the mean flow or from buoyancy at a scale not too far removed from that of the mean flow and is finally dissipated into heat by the smallest eddies of the system in which the velocity gradients are large enough for viscous effects to become significant. The statistics of the trubulence at low frequencies will be expected to be subject to strong directional bias, but, as the energy is progressively transformed into higher frequencies, there will be an increasing tendency toward equipartition among the components of the motion and an increasing lack of correlation between them. Thus, as the eddy scale diminishes, the eddies show fewer and ultimately no marks of their progenitors other than the energy drived from them. The domain of *local homogeneity* is said to obtain at points on the scale where the statistical properties depend only on the spatial separations and time differences involved in their formulation and *local isotropy* when there is local homogeneity and the statistical properties are independent of the orientation and sense of the axes. The smallest eddies should lie in this domain and can play no part in the directional transport of heat and momentum.

A number of theoretical relations pertaining to $r(\sigma)$, $F(n)$, and other statistical properties have been derived, principally by Russian and Japanese workers, based on similarity arguments. The majority apply to the isotropic range, for which the similarity arguments are most readily formulated, and so are of interest in the present context mainly insofar as experimental verification may indicate the limits of the isotropic range and hence the limits of frequency above which we do not need here to be so much concerned. Predictions from the similarity principle and their applications to meteorological problems in general have been reviewed by Priestley and Sheppard (1952) and MacCready (1953).

At the lower frequencies, where the greater interest attaches, the approach has, of necessity, been empiric, and the results are consequently inclined to be fragmentary. But a number of spectra and associated statistical functions have been evaluated, and the dependence of their characteristics on such factors as height, wind speed, and stability gives assurance that a worthwhile body of knowledge about the turbulent structure has been accumulated. We shall summarize these studies below, but, before doing so, a brief account of some results obtained from the atmosphere at the higher frequencies will be useful in indicating the total range of the turbulence and in identifying the ranges of operation of the different physical processes.

Viscous Dissipation and the Microscale of Turbulence

G. I. Taylor (1922) derived an expression for λ, the *microscale* of turbulence, which is a measure of the small eddies responsible for the viscous dissipation, as

$$\lambda^2 = \frac{10\nu E}{\epsilon}, \tag{5.4}$$

where E is $\frac{3}{2}\overline{u'^2}$, the kinetic energy per unit mass of the turbulence (assumed isotropic), ϵ the rate of viscous dissipation per unit mass, and ν the kinematic viscosity. It can be shown that

$$\epsilon = \frac{\tau}{\rho}\frac{\partial u}{\partial z}, \tag{5.5}$$

provided that we can neglect certain terms associated with the vertical transport of heat and of turbulent kinetic energy which are generally small by comparison, at least in the boundary layer. Thus R. J. Taylor (1952) was able to calculate λ from measurements of the shearing stress. The precise physical interpretation of λ is not, however, clear. Moreover, since the size of the smallest eddy should be determinable in terms of ν and ϵ only, there appear to be advantages in the alternative definition,

$$\lambda_2 = 15\left(\frac{\nu^3}{\epsilon}\right)^{1/4}. \tag{5.6}$$

The functional form is decided by purely dimensional arguments. The numerical multiplier has been determined from wind-tunnel experiments, within inevitably rather generous limits of accuracy, so that, according to MacCready (1953), λ_2 corresponds to the size of eddy below which 90 per cent of the viscous dissipation occurs.

Taylor's values of ϵ can quite readily be applied to give this alternative length, and MacCready calculated values of λ_2 from ϵ determined from the energy spectrum function according to the relationship derived as equation (5.7). Lettau (1954, 1957) has argued that equation (5.5) applies to much deeper layers, provided that the right-hand side is interpreted as the scalar product of the horizontal stress and vertical wind-shear vectors. By making use of the Leipzig wind-profile data of Table 2 and inferring τ by the methods described at the end of chapter 3, ϵ and thence λ_2 can then be found at much greater heights. There are, accordingly, three sets of comparable values derived not merely from independent data but from quite different types of data. It is to be expected that λ_2 will depend on both height and wind speed, V, and that, other things being equal, $\epsilon \propto V^3$, whence $\lambda_2 \propto V^{-3/4}$, smaller eddies being maintained against viscosity when more turbulent energy is available to feed them. Using the last relation to reduce all values of λ_2 to a standard wind speed of 5 m sec^{-1}, we may obtain the height dependence of λ_2 and compare the results of different workers, as in Figure 15. Quite a concordant picture is obtained. The line drawn in the diagram, for which $\lambda_2 \propto z^{1/3}$, provides a very fair fit over the whole range (all the data were obtained in daytime). Values of ϵ corresponding to λ_2 with $\nu = 0.145$ cm^2 sec^{-1} are shown on the scale on the right-hand side of the diagram.

The Inertial Subrange and the Limits of Isotropy

Equation (5.6) is an example of the similarity argument, following Kolmogoroff, that all statistical characteristics of the range of smallest sizes must be functions of ν and ϵ only. The range for which this applies is known as the *viscous subrange*. It may well be that there also exists a range of sizes which are small enough to be isotropic, yet sufficiently large to be removed from any direct influence of viscosity. Here the character of the turbulence must be governed solely by the amount of energy passed down from the larger ranges

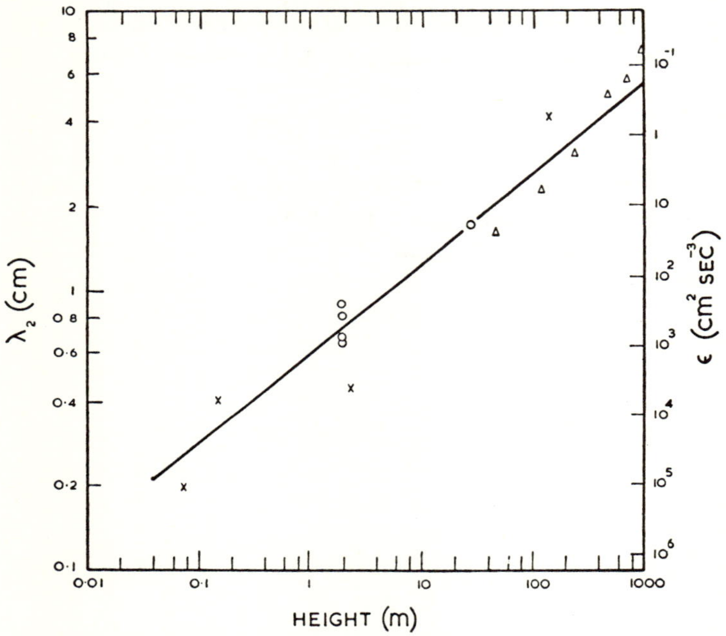

FIG. 15.—Microscale of turbulence and dissipation rate, for wind speed of 5 m sec^{-1}. × = MacCready; ○ = Taylor; △ = Lettau.

of eddies and in turn passed on to the smaller ranges, so that the statistical characteristics within this range can be functions of ϵ only. This is known as the *inertial subrange*, and its boundary on the low-frequency (large-eddy) side marks the limit of isotropy. It is of some interest to determine whether the inertial subrange exists in the atmospheric boundary layer and where its limit lies.

In the inertial subrange the cross-correlation and co-spectrum functions are zero by definition, but the functional forms of the energy spectrum and autocorrelation functions, $F(n)$ and $r(\sigma)$, may be inferred by dimensional analysis from the appropriate similarity assumptions. The derivation is not unique, unless it is further assumed that time resolutions as measured at a fixed point are relevant only as implicit distance resolutions or, in other words, that frequency n appears only in association with $\bar{u}$ as a wave number and σ only in association with $\bar{u}$ as a distance. The assumption appears to be based on the picture of a constant pattern of turbulence swept forward at the mean wind speed and is likely to be more nearly valid in the wind tunnel than in the atmosphere, where $|u'|/\bar{u}$ is typically larger. It then follows that

$$F(n) \propto \epsilon^{2/3} n^{-5/3}\,, \tag{5.7}$$

$$[1 - r(\sigma)] \propto \epsilon^{2/3} \sigma^{2/3}\,. \tag{5.8}$$

The constants of proportionality have been discussed by MacCready, among others.

MacCready (1953) and Taylor (1955) have tested equation (5.8) with almost identical results. In five tests at heights from 7.5 cm. to 146 meters, MacCready found that curves of the shape specified will fit the observed autocorrelation function up to a certain value of σ, but the test was not a discriminating one, as other curves would have fitted equally well. Taylor made twenty-five tests at heights of from 1.5 to 29 meters, measuring in each case on a logarithmic plot the value of the index of σ in equation (5.8) which best fitted the data at the smaller values of σ. The mean index, with its standard error, was 0.69 ± 0.04. Both authors found, however, that in nearly every case the fitted relation extended to a value of σ about equal to z/u and generally well beyond this point. Eddies of this size must be influenced by the proximity of the ground, and so would not be expected to be isotropic. By applying other tests, Taylor has shown that, in fact, there was not isotropy; equation (5.8) is a necessary, but not a sufficient, condition, and there may well be other postulates than those of the inertial subrange on the basis of which equation (5.8) is valid. The autocorrelation function has no precise resolving power with respect to eddy size or frequency; as equation (5.3) shows, the value of $r(\sigma)$ is affected by eddies over the whole range of sizes, both smaller and greater than σ. The limits of validity of equation (5.8) may not therefore correspond very closely with the limits of frequency of the isotropic range. The reader is reminded of the warning about physical interpretation with which the mathematical functions were introduced.

The spectrum functions should provide the more distinct resolution with respect to size or frequency. MacCready has verified the form (5.7) for the energy of the u, v, and w components over the range 2.5–40 cycles per second in somewhat unstable conditions, at heights of 1 and 2.4 meters in winds of 2 and 4.6 m sec^{-1}, respectively. With an eye to comparison and generalization, the result is conveniently stated as applying to the range 1.2–20 of the non-dimensional nz/u. The result may then be related to those in Figure 3 by noting that, broadly speaking, the use of a moving average of period t_m will retain most of the contribution from frequencies lower than $1/2t_m$ and suppress most of that from higher frequencies. MacCready's result would accordingly imply no loss of flux in Figure 3 for values of $t_m u/z$ less than $1/2.4 \approx 0.4$. Thus there is good evidence for placing the isotropic limit between this value and $t_m u/z = 0.8$, corresponding to $nz/u = 0.6$, beyond which contributions to heat and momentum flux are definitely indicated in the diagram. Panofsky (1953*b*) has evaluated the stress spectrum, or co-spectrum of u and w, obtained for $z = 22$ meters, $u = 3$ m sec^{-1}, and four other examples at $z = 91$ meters with u from 3.5 to 10 m sec^{-1}, all in superadiabatic conditions. The high-frequency ends are generally well defined and within about 25 per cent of $nz/u = 0.6$ in every case. As with λ_2, the results of different workers converge to identify, within acceptable limits, a characteristic point of the spectrum and include the means of generalizing over a fair range of heights and wind speeds.

Peaks of Energy and Co-spectra

The low-frequency end of the range, where the turbulence is no longer isotropic, must embrace the whole of the co-spectra and must also include the peaks of the energy spectra, since (5.7) is a decreasing function of n. Moreover, it is here that the energy enters the system and the influence of stability will be most directly felt.

The spectral distributions within this region have been studied most intensively by Panofsky and his collaborators, and many of their results are summarized by Panofsky and McCormick (1954).

Twenty spectra have been worked out under daytime conditions at heights mainly around 100 meters and the results averaged in groups *a*, *b*, and *c* according as the incoming radiation was, respectively, large (≥ 1 cal cm^{-2} min^{-1}), intermediate, or, for practical purposes, negligible (≤ 0.2), as these last occasions were, under the conditions of the trials, those of strongest wind. While a more dynamic parameter such as *Ri* would have been preferable to the one selected, classification in terms of radiation is adequate to bring out

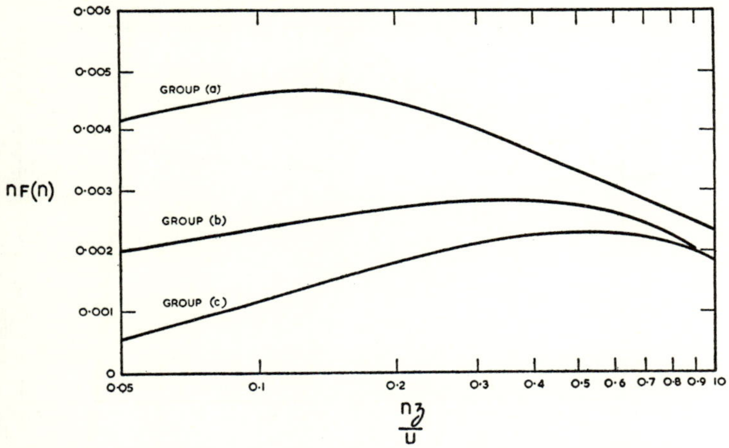

FIG. 16.—Spectra of *w*-fluctuations as a function of normalized frequency (Panofsky and McCormick).

many of the more significant trends in behavior. For $nz/u \geq 0.6$ it was found that the intensity of *w* fluctuations, and even more closely those of *u*, are almost independent of radiation (though it should be noted that inversion data were not included). At lower frequencies there was generally considerably more energy in the horizontal than in the vertical component, though this did not remain true in group *a*, *w* being shown to be much more strongly dependent on radiation than *u*, the ratio $F_{ww}(n)/F_{uu}(n)$ increasing steadily with radiation. As the radiation increases, there is not only a total increase in the energy in the low frequencies but a progressive shift in the peak of the spectrum toward lower frequency. This is illustrated in Figure 16, where the average *w* spectrum for each group is shown as a plot of $nF(n)$ against n, the latter on a logarithmic scale so that the area under the curve represents the energy amount. The value of n has been normalized prior to the averaging by multiplication by z/u. But the double averaging used in the production of Figure 16 has greatly smoothed out the sharpness of the peaks as they appear in the individual spectra and has lost altogether at least one other highly significant feature. Panofsky himself was the first to remark that the typical individual spectrum in strongly unstable conditions tended to have not a single but a double peak, as is indeed quite

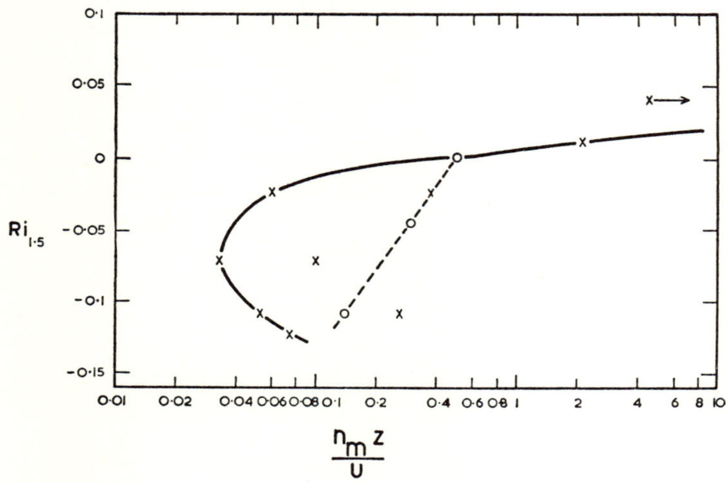

FIG. 17.—Frequency of w spectrum peaks as dependent on stability. × = Webb; ○ = Panofsky and McCormick (estimated).

strongly suggested from even a visual inspection of the original records, such as the upper curve of Figure 14.

The occurrence, in unstable conditions only, of the double peak in the spectra of both the w and the u component has been fully verified by Webb (1955). The particular interest in Webb's series of spectra is that they cover a wide range of stabilities, including moderately strong inversions, at a constant height (29 meters) and mainly in a narrow range of wind speeds (5.3–7 m sec^{-1}). Figure 17 shows the normalized values of n_m for the w spectrum plotted against $Ri_{1.5}$, n_m being the frequency of the peak of $nF(n)$. The corresponding points obtained from Figure 16 are also indicated, though it is not, of course, possible to do better than estimate their ordinates rather roughly. The full curve has been drawn quite uncritically to pass through all the low-frequency peaks, these containing in every case the greater part of the kinetic energy. It appears that Webb's peaks are at a consistently lower normalized frequency than those of Panofsky and McCormick. This may be due to the rather crude averaging of Figure 16. However, it must be noted that Webb's height was much the lower; to the extent that the energy is dominated by deeply penetrating convective elements, it is the frequency itself, rather than nz/u, which should remain constant with height on a given occasion. Panofsky and McCormick's own more detailed comparisons do, in fact, show a tendency for $n_m z/u$ to increase with height, though not so rapidly as constancy of n_m would imply (see Table 4), and the conclusion must be drawn that the energy is strongly influenced, but not dominated, by these deep penetrations.

The study of what is known as the *quadrature spectrum*, based on the lag characteristics between u and w fluctuations, also enables deductions to be made about the shape of the eddies and their disposition relative to the height of the observing point. Panofsky and McCormick conclude that the low-frequency eddies in strong radiation are taller relative to their breadth than at other times.

Another point of interest in Figure 17 is that the peak of the w spectrum in inversion conditions lies at much higher frequency than the value $nz/u = 0.6$, which has been indicated as marking the isotropic limit in the unstable case. On the other hand, Webb found that the peaks of the u spectrum were less than 0.6 on each of three inversion occasions with $Ri_{1.5}$ up to $+0.1$. The influence of stable stratification is therefore to displace the isotropic limit much further toward higher frequencies and in the low-frequency ranges to suppress the energy in vertical, relative to that in longitudinal, motion. A similar effect as between the w and the v (cross-wind) fluctuations has long been known from bidirectional vane studies.

One last conclusion from the spectral analyses has been the full verification of the effect illustrated in Table 4, that the stress spectrum shows a consistent shift from the w energy spectrum toward the lower frequencies. Panofsky and McCormick find some support, both in the observations themselves and in a theoretical generalization of the mixing length concepts, for a relation of the form

$$F_{uw}(n) \propto n^{-1} F_{ww}(n) .$$

The limitations of the theory and the scatter of the observations leave room for alternative forms, but the result is useful as a general guide.

The Structure of Free Convection

Because of the general prevalence of the idea that free convection, when it occurs, will take the form of regular cellular structures of the Benard type, it seems desirable to examine the argument on which the conception is founded. According to the classical treatments of Rayleigh, Jeffreys, and others, such cells can form in still air in a layer of depth h only if

$$-\frac{(g/\theta)(\partial\theta/\partial z)\,h^4}{\kappa\nu} > 1{,}100 \text{ at least}.$$

To the extent that any such criterion could be expected to apply when there is pre-existing wind and turbulence, it would appear suitable to replace the molecular conductivity, κ, and kinematic viscosity, ν, by their turbulent counterparts,

$$K_H \approx K_M \approx k^2 z^2 \left(\frac{\partial u}{\partial z}\right).$$

These vary with height, but their average value through the depth h of the layer would be given by

$$\overline{K_H} \approx \overline{K_M} \approx a k^2 h^2 \,\overline{\frac{\partial u}{\partial z}},$$

where a is less than, but comparable with, unity. In this way the Rayleigh criterion would be transformed into a Richardson criterion,

$$-\frac{(g/\theta)(\partial\theta/\partial z)}{(\partial u/\partial z)^2} > 28a^2.$$

The disparity between this result and the evidence that a free-convection regime sets in at about $Ri = -0.03$ suggests most forcibly that the criterion as derived above is not the relevant one when we are considering the onset of free convection under conditions of some wind, which is, of course, the normal state in nature. Whether this implies that regular cells will not form is less certain because of the lack of rigor in the above argument, but the view is supported by the fact that the examination of numbers of records taken in such conditions has failed to detect any marked regularities.

From the preceding discussion of spectral analyses it would be expected that free convection would be accompanied by a strong concentration of heat flux toward the low-frequency limit of the w spectrum. A clear but characteristic example is shown in Figure 18, which is an excerpt from an Edithvale record of simultaneous values of w, T, and the product wT' taken at a height of 23 meters, with

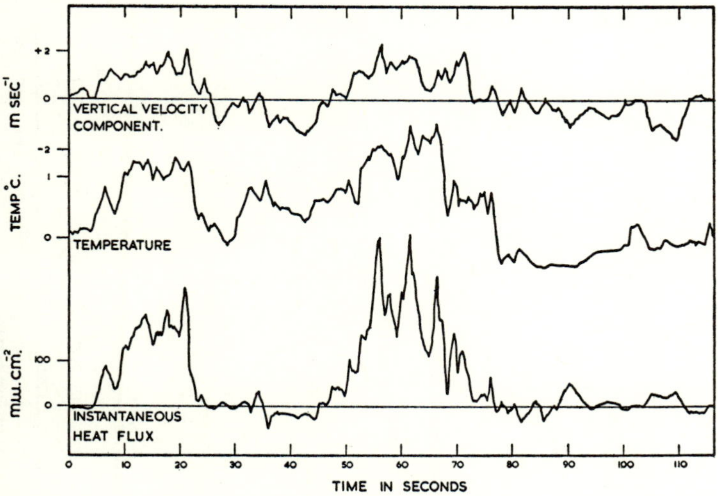

FIG. 18.—Association of temperature fluctuations with vertical velocity and heat flux.

$u_2 = 4.6$ m sec^{-1} and $Ri_{1.5} = -0.07$. The association of the larger temperature fluctuations with the broader, slower fluctuations of w and their combined dominance of the heat flux requires little further emphasis. Certain details of the structure may be even more clearly brought out in Figure 19, showing T, inclination i of motion to horizontal, and total horizontal speed, V, at a height of 7 meters in a lighter wind (1.6 m sec^{-1}) with $Ri_{1.5} = -0.18$. Here disturbed periods in the temperature trace, which are seen to be associated with ascending air, alternate with well-defined and almost fluctuation-free periods in descending air. This trace is part of a considerably longer period throughout which the minimum temperature re-

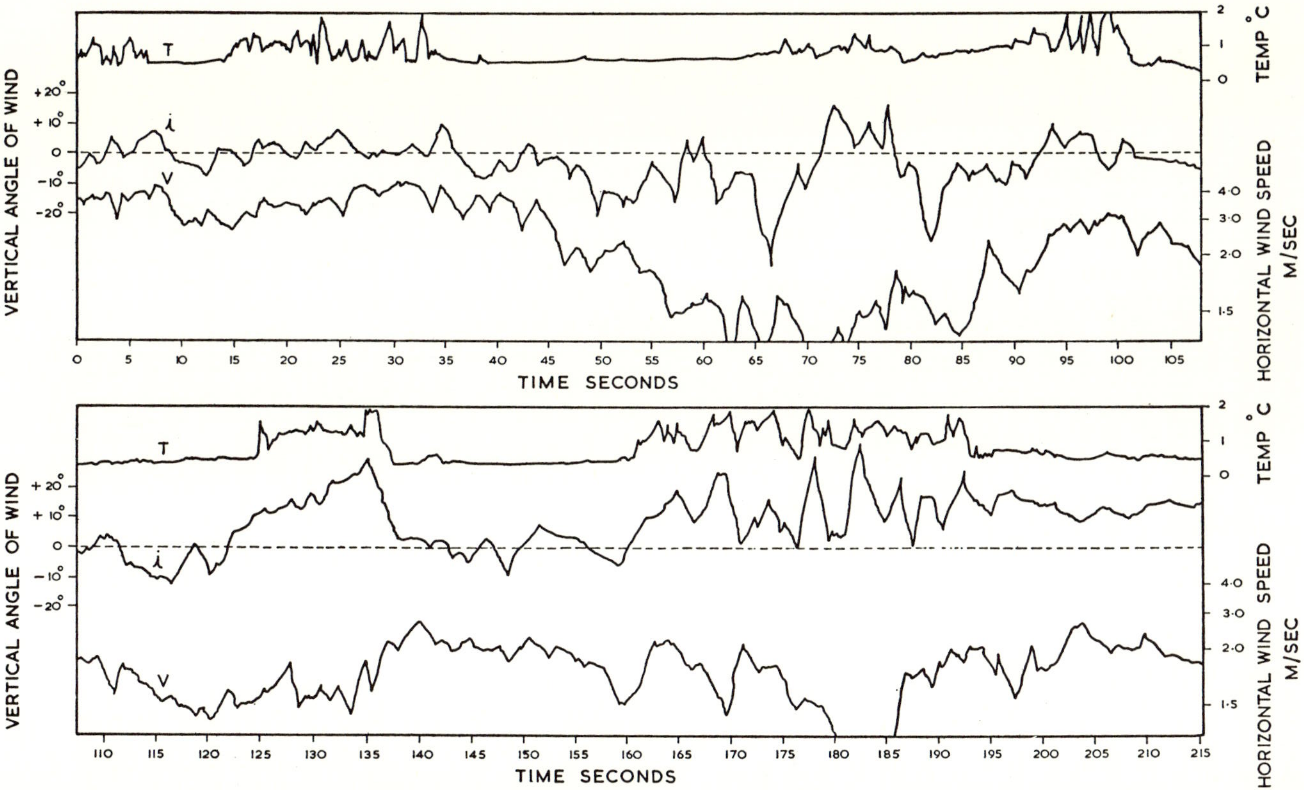

FIG. 19.—T, i, and V fluctuations at 7 meters ($Ri_{1.5} = -0.18$)

mained constant from one quiet interval to another, while the maximum temperature of the disturbed periods remained equally constant. There is, be it noted, no cessation of turbulence during the quiet intervals, and the smaller eddies, as judged from the V and i traces, retain much the same structure as they did during the disturbed intervals. The most likely explanation is that the sampling level is in, or sufficiently close to, an adiabatic layer of the type discussed in chapter 4, so that there is little temperature fluctuation in generally downward-moving currents. On the other hand, it may be judged from the relative constancy of the height of the peaks in the disturbed periods that there is a considerable frequency of motions originating from close to the surface and carrying the full temperature contrast which the mixing en route will allow. The complete contrast in character between the T and V traces is an example of the detailed individual evidence underlying the statistical evidence of chapter 2 for a difference in mechanism between heat and momentum transfer.

No further proof is needed that the lower-frequency disturbances characteristic of this state are distinguished for their abnormal depth of penetration, but it is of interest in Figure 20 to see the consequences when simultaneous temperature records are taken at several heights: on this occasion $u_{1.5}$ was 3.8 m sec^{-1}, $Ri_{1.5} = -0.05$. Here, as in Figure 18, is to be noted a marked asymmetry in the temperature disturbances, with the gradual ascent to the maximum seen in contrast to the sharp descent on the other side. The features of the individual bursts are quite distinctive, allowing the advent of the same burst at different heights on the tower to be identified, as by the broken lines. The order in which they arrive and even the fact that they do arrive at all levels on a single tower when the vertical velocity (from 30 to 80 cm sec^{-1} according to height) is so much smaller than the horizontal may at first cause some surprise. Taylor (1958*b*) has taken a number of records of this sort and shown that both effects are quite typical, and, with this as evidence, we may discuss the implications as to structure. There can be no doubt, from the appearance of the individual records such as Figures 18 and 19 or the simple fact of free convection, that the elements marked in Figure 20 represent rising elements of air whose motion originates below the lowest level of observation, since positive correlation between *w* and *T* already exists there. An element which "originates" suddenly and locally—i.e., a bubble—would not normally impinge on all levels of the tower, and the same may be said of a plume or continuous stream of bubbles coming from some stationary point to windward. If the source is stationary, the plume will occupy a fixed position in space, impinging on the tower at some levels only. If the

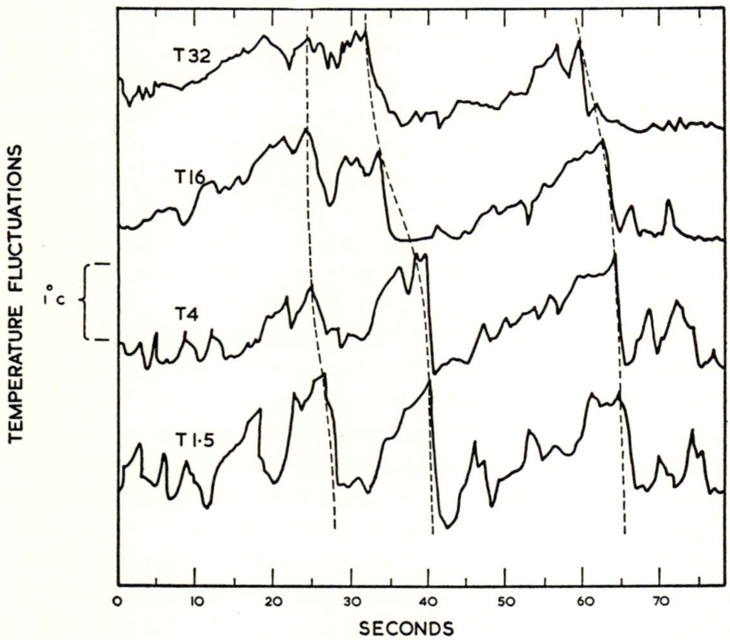

FIG. 20.—Simultaneous temperature fluctuations at 1.5, 4, 16, and 32 meters

source is now moved upwind or downwind at a steady speed, the plume will move as an entity with the same speed. The motion has two effects: it will affect the apparent shape of the plume as viewed by an observer at the side and, if continued for long enough, will insure that the plume impinges on all levels of the tower. The sequence of impacts, however, will now depend on the relative motion of the source and that of the wind in the layer concerned. If the source moves downwind faster than the wind, or upwind at any speed, the plume will trail behind the source, and the sequence of impacts will be progressively upward. Only if the source is moving downwind more slowly than the wind will the plume incline forward, as the observed sequence of impacts implies. In the example illustrated, the source must have traveled at least 300 meters.

It is established, then, that the convective elements are of the continuing plume type originating at, and moving with the speed of the wind at, some quite low level close to or below the lowest height of the observations (1.5 meters). This picture is supported by the characteristic unsymmetrical shape of the T anomalies, since the hottest elements of the plume would acquire the greatest upward velocities and so be found close to the upwind side when a cross-section is taken through any level. It seems reasonable to suppose that the disturbance, being a continuing one, will move with the speed of the wind at the level which "causes" the convection and therefore that this level is a very low one. Beyond this it does not seem possible to distinguish whether this cause is more immediately of a thermal nature, such as a hot spot on the ground, or is due to mechanical stirring or other internal influences which bring about temperature anomalies within the medium, as originally suggested by Priestley and Swinbank (1947). Recognition that forced convection will dominate at a sufficiently low level, provided that any wind at all is present, may incline the view toward the mechanical origin, but probably both processes are operative to some degree, and the desire to discriminate is more academic than practical.

The bending-back of the curve in Figure 17 at extreme negative Ri now admits a ready physical explanation. The normalized frequency is inversely proportional to the product of period and wind speed and so behaves as the reciprocal of what has been called the horizontal *scale length* of the eddies. Since Ri is more sensitive to wind shear than to temperature gradient, the phenomenon invites interpretation as a tendency for the plumes to become more or less drawn out along the wind direction according to the strength of the wind. To explain the effect is not, of course, to demonstrate its reality, nor are Webb's results in themselves sufficient to do so. However, we are reminded of the beautiful observations by Woodcock (1940) on the flight habits of herring gulls, which did not soar, soared linearly, or soared in circles at what quite clearly are progressively increasing magnitudes of the negative Richardson number.

On the question of the actual magnitude of the w and T fluctuations and their variation with height and lapse rate, the dimensional analysis of chapter 4 which led to equations (4.4)–(4.6) can again be applied, to give

$$\sigma_w = a \left(\frac{gC}{T}\right)^{1/2} z^{1/3}, \tag{5.9}$$

$$\sigma_T = bC\, z^{-1/3}, \tag{5.10}$$

where a and b are constants whose theoretical values, like that of $\overset{*}{H}$, will depend upon the details of the convection model. It has been found (Priestley, 1956*b*) that equation (5.10) is obeyed over a wide range of negative Ri with b about 1.2 but that equation (5.9), with $a \approx 1$, holds only at the most extreme conditions; this reflects the evidence in the figures that σ_w retains a significant contribution from mechanically generated turbulence, while σ_T is the more dominated by the free-convection structure.

Before leaving the discussion of structure, it seems worthwhile to point to the strong resemblance between some of the diagrams of this chapter and what is observed in larger convection on the scale of cumulus clouds, especially to that between Figure 19 and the alternation of cumulus-full and cumulus-free patches in the trades described by Malkus (1957). For example, if l be used to denote a characteristic length of the convection element, we may compare typical values as given by Swinbank (1955) and Taylor (1956*a*) with those obtained from the largest cumulus (Byers and Braham, 1949), and it will be found that the dimensionless ratios l/z, $l/\theta\ \partial\theta/\partial z$, $l/w\ \partial V/\partial z$, $w^2\theta/gl\theta'$ are of the same magnitude on the two scales. There will, of course, be differences of detailed behavior due to processes of change of phase, but a strong measure of dynamical similarity exists in the broad pattern, and the study of each scale should be of concern to the other.

6. THEORIES OF BUOYANT MOTION

Principal Types of Buoyant Motion

It has been seen that our present knowledge of the effects of stability on the turbulent transfer of momentum is entirely empirical, while that of heat transfer retains strong elements of empiricism. In order to achieve success in any more fundamental approach, the motion of an individual element under the influence of buoyancy must come under particular study. This is an aspect which has not been integrated in previous texts. And, since it forms one of the strongest links between boundary-layer motions and those involving much larger depths of atmosphere, the present chapter may appropriately be set in a somewhat wider context than that of the remainder of the volume. In instances where the most obvious application appears to be directed toward the larger scale, it must be borne in mind that the equations and principles involved are not specific as to scale and the existence of counterpart motions in the boundary layer with a strong measure of dynamic similarity is often to be inferred.

The classical mixing length theory was based on the concept of randomly moving *elements* whose size and configuration did not need to be precisely defined, beyond that they represent associations of neighboring particles which have in common a measure of velocity, temperature, humidity, etc., which may be different from that of the environment. The concept differs not at all from that of the *parcel*, which has been an important tool in the study of more purely convective motions, but there has been a difference in treatment between the two fields of inquiry, insofar as the motive force of buoyancy has generally been taken into account in the latter case but not in the former. In an attempt to reconcile the two and to introduce into both the consideration that the mixing between the element or parcel and its environment will be a gradual, rather than an abrupt, process, a formulation was given by Priestley (1953); the object is to define the principal types of elemental motion which will occur under the motive force of buoyancy and the environmental conditions appropriate to each, without insisting more than is necessary at this stage on the details of the structure of the element itself.

Using suffix e to denote conditions in the environment and primed symbols for excess of element over the environment, we may write the basic equations as follows:

$$\dot{w} = \frac{g}{T_e} T' - k_1 w \,, \tag{6.1}$$

$$\dot{T} = -w\Gamma - k_2 T' \,, \tag{6.2}$$

where k_1 and k_2 are mixing rates for momentum and heat and the environment is assumed to have no mean vertical motion. The condition for a steady environment is

$$\dot{T}_e = w \frac{\partial T_e}{\partial z} .$$

Assuming that the environmental lapse rate is constant and that the mixing rates remain constant for a given element, we may eliminate one of the variables w, T' from equations (6.1) and (6.2) and obtain the differential equation for the other. Neglecting a quadratic term which is, in general, small, the resulting equation for w is

$$\ddot{w} + (k_1 + k_2)\,\dot{w} + \left[\frac{g}{T_e}\left(\frac{\partial T_e}{\partial z} + \Gamma\right) + k_1 k_2\right] w = 0 \,. \tag{6.3}$$

This is, for practical purposes, an equation of the second order with constant coefficients. The details of the algebraic solution are straightforward, with three well-known basic modes of behavior dependent on the relative values of the coefficients; we shall here omit the algebra and give only the interpretation.

The treatment of the mixing rates as constant for a given element, rather than w-dependent, suggests that the formulation is more appropriate to an environment already in turbulent motion than to the case where the mixing is induced by the element's own motion. Priestley argued that, with a given intensity of turbulence, k_1 and k_2 can be regarded as inverse measures of the size of the element under consideration. It then follows as a general property of the solutions in both superadiabatic and subadiabatic conditions that small enough elements, assigned a finite initial velocity w_0 and/or temperature excess T_0', will move only a finite distance. The approach to the final position is asymptotic, and the total vertical distance moved, or buoyancy length, l_b, is given by

$$l_b = \frac{|(gT_0'/T_e) + k_2 w_0|}{(g/T_e)[(\partial T_e/\partial z) + \Gamma] + k_1 k_2}, \tag{6.4}$$

which is necessarily positive for the smallest elements and increases with increasing size of element, other things being equal. When the element exceeds a critical size, which becomes smaller as the lapse rate departs increasingly from adiabatic in either sense, a second and third mode of motion will obtain for subadiabatic and superadiabatic conditions, respectively. In the former case, equation (6.4) still holds, but the element first transcends a final equilibrium and subsequently executes damped harmonic oscillations with period

$$2\pi \left| \frac{g}{T_e}\left(\frac{\partial T_e}{\partial z} + \Gamma\right) - \frac{(k_1 - k_2)^2}{4} \right|^{-1/2} \tag{6.5}$$

In the latter case the vertical velocity eventually increases exponentially with time, i.e., a mode of true instability results.

The most novel feature is the recognition that two elements in the same environment may behave quite differently by virtue merely of their size, and the picture thereby derived is both more detailed and more realistic than the simple stability-instability alternative of the parcel model. The use of k constant with time means that progressive growth of the element is not allowed for, and this is a weakness of the theory. Nevertheless an advance has been made from the classical model, and the general relevance of the results will be apparent.

On this foundation it was possible in a subsequent paper (Priestley, 1954*b*) to make some theoretical analysis of the relative contributions to the heat flux of elements of different sizes. It was supposed that a population of elements of all sizes was set in motion initially by impressing temperature anomalies of a given magnitude; by taking statistical combinations of the appropriate solutions of equation (6.3), a relation was obtained between the upward heat flux, H, so resulting and σ_T^2, the variance of temperature at the same level. The relation depended on the lapse rate and mixing rates (element sizes) but was independent of the actual magnitude of the impressed anomaly. The chief result was to provide, for the first time, an indication of the magnitude of the upward heat flux which might be achieved through a layer of subadiabatic temperature gradient. It is obvious that the smallest eddies will then be inefficient as agents in heat transport, and it was shown that the largest elements will also be inefficient, their mode of motion approaching a simple harmonic form with w and T' a quarter-cycle out of phase. There exists a condition on the mixing (or on the size of element) for which H/σ_T^2 is a maximum. This is given by

$$k_1^2 = k_2^2 = \frac{1}{3}\frac{g}{T_e}\left(\frac{\partial T_e}{\partial z} + \Gamma\right)$$

and the maximum of H by

$$\frac{H}{\rho c_p} = \frac{\sigma_T^2}{\sqrt{(3T_e/g)\left[(\partial T_e/\partial z) + \Gamma\right]}}. \qquad (6.6)$$

Bunker (1956) made measurements by the eddy-flux technique from an aeroplane flying in a stably stratified air mass over warm sea between Rhode Island and Bermuda, showing that upward heat flow can occur under such conditions and that formula (6.6), while not finally accurate, appears relevant to the problem. Nine measurements of heat flux were taken with $(\partial T_e/\partial z) + \Gamma$ positive, $+3.0 \times 10^{-5}$ ° C. cm^{-1} on one occasion and between 0.5 and 1.1×10^{-5} on the others, the heights being from 150 to 500 meters and σ_T from 0.05 to 0.1° C. Heat fluxes averaging 5 mw cm^{-2} were measured as against an average value of 3 mw cm^{-2} from equation (6.6); in every case the heat flux was upward, and the variations in its magnitude correlated strongly with those in the theoretical formula.

Equations (6.1)–(6.3) and their solutions serve usefully to present, in one general picture, the main modes of buoyant motion and the conditions of environment and element which govern their occurrence. Precision may be gained, inevitably with some narrowing of the horizon, by being more specific as to the detailed structure and mechanics of the element with which we are dealing. Morton, Taylor, and Turner (1956) consider the behavior under stable stratification of an element which remains spherical and progressively adds to its mass by entrainment of air from the environment at a rate proportional to w and to the surface area of the element. This adds a third equation, that of continuity, to those for momentum and heat (or density deficiency), and a third variable in the form of R, the changing linear dimension of the element.

Morton *et al.* assume that initially the element has infinite upward velocity and density deficiency, $-\rho'$, the radius being zero in such a way that $-R^3\rho'$, proportional to the total buoyancy, is finite. The height of the final equilibrium is then shown to be proportional to

$$(-R^3\rho')^{1/4}_{\text{source}}\left(-\frac{\partial \rho_e}{\partial z}\right)^{-1/4}, \tag{6.7}$$

which has been verified by releasing light fluid in a still tank of heavier fluid in which there was a stable density gradient. The different types of mixing law lead to contrasting results, as between (6.4) and (6.7), but the patterns of final behavior are similar; Morton *et al.* find that the element will first overshoot its equilibrium and subsequently oscillate with decreasing amplitude and period,

$$2\pi\left(-\frac{g}{\rho}\frac{\partial \rho_e}{\partial z}\right)^{-1/2}, \tag{6.8}$$

comparing closely with (6.5). R increases in proportion to the distance traversed up to the maximum of the latter and thereafter at a slower, variable rate.

The bases of the two treatments just described draw attention to the question whether the mixing from motions already in the environment or that induced by the element's own motion will be the more important influence in restraining the upward velocity and temperature or density anomaly. The entrainment models allow only for the latter, and this is true of the treatments which follow, though the choice may be held to have been one of mathematical convenience rather than physical belief. Evidence such as that of Figure 14 (p. 54) is a reminder of the limitation imposed by this neglect.

Theory of Buoyant Plumes

We come now to study one particular type of buoyant element which has been shown in chapter 5 to be especially relevant to the atmospheric boundary layer, in one condition at least, and has a yet wider importance in convective problems under all conditions—the plume or continuing convective element. The formal treatment of convection plumes has been developed by several workers, but the generalization to the thermally stratified environment, which is an essential consideration here, is due to Priestley and Ball (1955) and Morton, Taylor, and Turner (1956).

Priestley and Ball consider an axially symmetric plume in which r is radial distance and z is height, whence the equations of continuity, vertical motion, and heat conservation for an annular element are, respectively,

$$\frac{\partial}{\partial z}(rw\rho) + \frac{\partial}{\partial r}(ru\rho) = 0\,, \tag{6.9}$$

$$\frac{\partial}{\partial z}(rw^2\rho) + \frac{\partial}{\partial r}(ruw\rho) = r\frac{\theta'}{\theta_e}\rho g + \frac{\partial}{\partial r}(r\tau)\,, \tag{6.10}$$

and, with θ denoting potential temperature,

$$\frac{\partial}{\partial z}(rw\theta\rho) + \frac{\partial}{\partial r}(ru\theta\rho) = -\frac{1}{c_p}\frac{\partial}{\partial r}(rF)\,, \tag{6.11}$$

where u is radial velocity, τ vertical turbulent shearing stress, and F radial potential turbulent heat flux. A dependent kinetic-energy equation can be derived from equations (6.9) and (6.10) as

$$\frac{\partial}{\partial z}(\tfrac{1}{2}rw^3\rho) + \frac{\partial}{\partial r}(\tfrac{1}{2}ruw^2\rho) = rw\frac{\theta'}{\theta_e}\rho g + w\frac{\partial}{\partial r}(r\tau)\,. \tag{6.12}$$

The entrainment rate, $r\rho u$, is assumed to remain finite as r tends to infinity, but rF, $r\tau$, w, and θ' all tend to zero. The similarity hypothesis is then introduced as

$$\frac{w}{w_m} = f\left(\frac{r}{R}\right), \qquad \frac{\theta'}{\theta'_m} = h\left(\frac{r}{R}\right), \qquad \frac{\tau}{\frac{1}{2}\rho w_m^2} = j\left(\frac{r}{R}\right),$$

independent of z, where the suffix m denotes values on the axis, R a characteristic radius at height z, and f, h, and j functions which represent the profile shapes. Integration with respect to r from 0 to ∞ then leads to

$$\frac{d}{dz}(R^2w_m^2) = aR^2\frac{\theta'_m}{\theta_e}g\,, \tag{6.13}$$

$$\frac{d}{dz}(R^2 w_m^3) = bR^2 w_m \frac{\theta'_m}{\theta_e} g - cRw_m^3 , \tag{6.14}$$

$$\frac{d}{dz}(R^2 w_m \theta'_m) = -dR^2 w_m \frac{\partial \theta_e}{\partial z} , \tag{6.15}$$

as the three basic equations in the unknowns R, w_m, θ'_m; and a, b, c, and d are numbers which may be referred to as *profile constants*, since they are determined by the functions f, h, and j. Following evidence from laboratory studies by Schmidt (1941), Rouse, Yih, and Humphreys (1952), and Railston (1954) that a Gaussian curve would fit reasonably well to the profiles of both w and θ' and that the dispersion in the two cases was not widely different,[1] it was further assumed that

$$f = h = \exp\left(-\frac{r^2}{2R^2}\right),$$

whence $a = d = 2$, $b = 3$. With these numerical values, one part of the solution follows immediately and quite generally as

$$\frac{dR}{dz} = c , \tag{6.16}$$

so that the plume spreads linearly with height. The spreading coefficient c is dependent solely on the form of j and so can be said to be determined by the magnitude of the quadratic drag force. In an appendix, Priestley and Ball discuss the case in which the profiles are not quite normal and of unequal dispersion.

By using density deficiency instead of potential temperature, results of applicability to all liquids can be obtained. Morton *et al.* proceeded in this manner and replaced the quadratic law of resistance by the assumption of physical entrainment at a rate proportional to w and the surface area, i.e.,

$$\frac{d}{dz}(R^2 w_m) = 2\,\alpha R w_m . \tag{6.17}$$

In contrast, it follows from equations (6.13) and (6.14) that

$$\frac{d}{dz}(R^2 w_m) = 2\,cRw_m + R^2 \frac{dw}{dz} \tag{6.18}$$

1. The evidence on this point is confusing, the fitted curves of Rouse *et al.* showing somewhat greater lateral spread of heat in the axially symmetric plume but of momentum in the one-dimensional (linear) plume.

(numerical values $a = 2$, $b = 3$ have been inserted), which recognizes that entrainment in a plume should be acceleration-dependent as well as velocity-dependent. In other respects the two formulations are identical.

A number of variations of this basic treatment are possible, and the result is generally a slight variation in the form of equations (6.13)–(6.15) or a different set of numerical values of the profile constants. For example, the analysis may readily be extended from circular to elliptical symmetry, and, with the Gaussian assumption, the same set results, with $(R_1R_2)^{1/2}$ replacing R and the measures of dispersion, R_1, R_2, in the principal directions each increasing linearly with height. This, together with the consideration that the principles used apply irrespective of whether a wind is blowing, led Priestley (1956*a*) to suggest that the solutions could be adopted as working solutions in the problem of the bent-over plume with c, now $(c_1c_2)^{1/2}$, the chief wind-dependent quantity. The results obtained agreed rather well with measurements of plumes from factory chimneys by Bosanquet *et al.* (1950) and with plume ascents from smoke pots over much shorter distances (Ball, 1958), but there is a general lack of good experimental data in other than near-neutral conditions.

Proceeding to the extreme, a parallel analysis of the two-dimensional problem of the linear plume leads to equations identical with (6.13)–(6.15) but with R replaced by L, the measure of lateral dispersion, and its index reduced by 1 in every term. For the Gaussian profiles in this case $a = d = \sqrt{2}$, $b = \sqrt{6}$.

Morton (1957) has applied the approach to cumulus convection, introducing an additional equation for the conservation of water, an additional similarity assumption, and appropriate adjustments in the other equations for the effects of density and condensation. The details do not concern us here, but it is of interest to note the use of the "top-hat" profile, wherein f and h fall abruptly from unity to zero at a certain distance. This may more appropriately represent the rather sharp boundary of cumulus due to evaporation. For the top-hat profile $a = d = 1$, $b = 2$.

Alternatively, F. H. Schmidt (1957*a*, *b*) has proposed the form

$$\frac{w}{w_m} = \left\{1 - \tfrac{1}{2}k\left(\frac{r}{R}\right)^k\right\} e^{-(r/R)^k},$$

which with $k = 1.8$ fits the data of Rouse *et al.* tolerably well almost to $r = R$ and provides for a downward velocity from a short distance beyond. The basis for adopting this form is that it is consistent, through the equation of continuity, with ru becoming zero both at $r = 0$ and at $r = \infty$, or, in other words, that there shall be no entrainment at sufficiently large distances. Equations (6.13)–(6.15) must be modified accordingly and become rather less tractable. Schmidt's analysis has much merit for the case of plumes in a closed chamber, but in the real atmosphere it appears doubtful whether a plume is immediately surrounded by its compensating downdraft except in cumuli where the effect may be attributed to other causes, and the retention of the concept of the plume as an entraining entity would appear to have advantages. This is certainly so where we are concerned with a close population of plumes in layers near the ground, where there is no evidence of any regular association between individual up-and-down currents, and in what follows we shall retain the models which include entrainment.

We shall next briefly review the individual solutions of plume motion in any thermal stratification, which constitute a requisite part of the general problem of vertical transport. So far, however, the necessary statistical application to give specific information about the flux has been attempted in only one special case, to be discussed in the final section. One aspect which has emerged from the foregoing discussions is the area of ignorance which surrounds the exact nature of the entrainment or mixing law and the appropriate distribution of w and T' within the rising element. These are not disconnected questions, as Schmidt has shown, and not purely hydrodynamic ones, as will become evident; their clarification by proper recourse to the principles of thermodynamics and basic turbulence theory, not neglecting the influence of pre-existing turbulence in the most general atmospheric case, is an outstanding theoretical task.

Properties of Solutions in Constant Lapse Rate

Under neutral conditions it follows from equation (6.15) that

$$R^2 w_m \theta'_m = \text{Constant } A \, , \qquad (6.19)$$

so that then and only then is the vertical heat flux in the plume constant with height and equal to the strength of the source. With the single restriction that $3a = 2b$, which is satisfied for the equal Gaussian assumption and for an infinity of other profiles, a complete solution is obtained in the form

$$R = cz\,, \tag{6.20}$$

$$w = \left\{\frac{3Ag}{2\theta_e c^2 z} + \frac{B}{z^3}\right\}^{1/3}, \tag{6.21}$$

$$\theta' = \frac{A}{c^2 z^2}\left\{\frac{3Ag}{2\theta_e c^2 z} + \frac{B}{z^3}\right\}^{-1/3}, \tag{6.22}$$

where B is a second constant of integration and the third has been fixed in equation (6.20) by positioning the origin of z, which was previously arbitrary, at the point of concurrence of the outline of the plume. This coincides with the source only in the case of an idealized point source, but for finite sources of radius R_0 it lies at a distance $z_0 = R_0/c$ below the source. The same solution has recently been obtained by F. H. Schmidt (1957*a*) for the non-entraining plume, also with a single restriction on the disposable constants of the problem. It includes, for $A = 0$, the solution obtained by Tollmien (1926) for a projected jet without buoyancy and also, for $B = 0$, the first special solution for the buoyant plume obtained by W. Schmidt (1941). But it is more than a mere combination of these two solutions because it allows arbitrary finite values of R, w, and θ' at the source to be incorporated and includes the meteorologically important case of a plume which accelerates upward at the lower levels and so develops vertical momentum from its initial buoyancy. All other analytical solutions appear to have been exclusively concerned with the decadent stage.

For discussing the case of thermal stratification we shall return to the equal-normal distribution assumed by Priestley and Ball and by Morton *et al.*, though some of the ensuing results apply more generally. It is seen that equation (6.20) still follows as a deduction from the premise (not as an additional assumption as some commentators have suggested) but that equations (6.21) and (6.22) no longer apply, while for equation (6.19) we obtain the more general integral,

$$R^2 w\left(\theta'^2 + \frac{\theta_e}{g}\frac{\partial \theta_e}{\partial z}w^2\right)^{1/2} = \text{Constant } A\,. \tag{6.23}$$

Here $\pi \rho c_p A$ is still equal to the source strength, Q, provided that w is small initially.

Under stable stratifications, for all but very weak sources, Priestley and Ball were able to show that the solution then had the following properties. Near to the source, the behavior closely resembled the solution represented by equations (6.21) and (6.22), or, in other words, the thermal stratification was of little account in the initial stages. Thereafter, it became more important, the plume ascending to a level given by

$$z_b = 1.10\, c^{-1/2} \left(\frac{g}{\theta_e} \frac{\partial \theta_e}{\partial z}\right)^{-3/8} \left(\frac{A g}{\theta_e}\right)^{1/4}, \tag{6.24}$$

where the temperature became equal to that of the environment. Above this level the solution showed some residual upward momentum operating against progressively increasing negative buoyancy until the motion was brought to rest at a *ceiling height* given by $z_c = 1.41\, z_b$. The best value of c was determined as 0.09 from rates of spreading in laboratory experiments.

Morton, Taylor, and Turner assume a source for which R tends to zero while w and $\Delta\rho$, the density deficiency, tend to infinity. With the single change in the equations noted above, they find the reversal of buoyancy at

$$z_b = 0.98\, \alpha^{-1/2} \left(-\frac{g}{\rho} \frac{\partial \rho_e}{\partial z}\right)^{-3/8} \left(R^2 w g \frac{\Delta \rho}{\rho}\right)^{1/4}_{\text{source}}, \tag{6.25}$$

with $\alpha = 0.09$ also obtained from the laboratory, and the ceiling at $z_c = 1.32 z_b$. The outline of the plume in this solution increases approximately linearly with height up to z_b and much more rapidly thereafter.

The occurrence of source strength to the $\frac{1}{4}$ power emphasizes how greatly the source strength must be changed to alter substantially the height of penetration, a point of some importance in the sphere of industrial pollution. To illustrate the magnitudes, Morton *et al.* gave a table for z_c for various typical sources in a calm I.C.A.N. atmosphere (lapse rate 6.5° C. km^{-1}). In Table 5 the corresponding values of z_b are reproduced: as it happens, equations (6.24) and (6.25) give virtually identical results. But z_b is preferred for reference, as both solutions come in conflict with the second law of thermodynamics above this level; they can be shown to imply that heat is transferred from a cold plume into a warmer environment.

Priestley and Ball ascribed this anomaly to the neglect of vertical diffusion, which clearly becomes most serious where θ is changing most rapidly. It may be noted that a formulation such as that of equations (6.1) and (6.2) must avoid the error, but, of those specific to plumes, only F. H. Schmidt's treatment has so far allowed for vertical diffusion, and this applies only to neutral conditions. An alternative way out of the difficulty would be to include a lateral heat-flux relationship of the form

TABLE 5

HEIGHTS OF NEUTRAL BUOYANCY z_b IN AN I.C.A.N. ATMOSPHERE

Typical Source	Q (Kw)	Height (Meters)
Household chimney..........	8	60
Bonfire.....................	450	150
Large power station chimney..	5×10^5	900
Burning town...............	2.5×10^7	2,400

$$\frac{F}{\frac{1}{2}\rho w_m \theta'_m} = k\left(\frac{r}{R}\right),$$

where k is another (positive) similarity profile function. This may be incorporated in an entropy equation, which can readily be derived from equations (6.9) and (6.11) as

$$\frac{\partial}{\partial z}(rw\rho \log\theta) + \frac{\partial}{\partial r}(ru\rho \log\theta) + \frac{\partial}{\partial r}\left(\frac{rF}{c_p\theta}\right) = -\frac{rF}{c_p\theta}\frac{\partial}{\partial r}(\log\theta)$$

which, when integrated across the plume, with θ'/θ_e small, yields

$$\frac{d}{dz}(R^2 w_m \theta'^2_m) + 3\frac{\partial\theta_e}{\partial z}\cdot R^2 w_m \theta'_m = -c'Rw_m\theta'^2_m, \quad (6.26)$$

c' being another positive constant dependent on the form of k. This together with equations (6.13)–(6.15) would then provide four basic equations in the three unknowns. The condition for consistency of the four equations would impose a single restraint on the values of the profile constants and prevent any violation of the second law, but the details of solutions by this approach have not been worked out.

For the case of the constant superadiabatic lapse rate, Priestley and Ball obtained a solution in the form of a hyperbolic integral which could be represented in the form of graphs of w and θ' against z for various sources and lapse rates. The initial behavior is again similar to that of the neutral solutions (6.21) and (6.22), and, under certain conditions of lapse rate and source strength, this similarity extends into a region of deceleration, though ultimately, provided that the superadiabatic layer is of sufficient depth, w and θ' increase together; θ' is everywhere positive, and there is no conflict with the second law.

The Free-Convection Regime

With the body of evidence that convective elements strongly resembling plumes are not only a feature of the free-convection regime but tend to dominate the heat flux under these conditions, it is of special interest to examine their behavior theoretically in the lapse-rate regime given by equation (4.5). From dimensional arguments, Batchelor (1954) has derived the indexes of the simple power-law solutions for this case. Priestley (1956*b*) gave full particular solutions of equations (6.13)–(6.15) and of the corresponding equations of the linear plume, using the equal-Gaussian profile assumption, applying these solutions to obtain theoretical heat-flux relationships, i.e., theoretical values of $\overset{*}{H}$, from the models. The derivation can be freed from any restrictive assumption about the profiles across the plumes and is reproduced below in its more general form.

Substituting for $\partial\theta_e/\partial z$ from equation (4.5) in equation (6.15), the solution is found as

$$R = \left(\frac{3a}{8b - 9a}\right) c\, z\,, \tag{6.27}$$

$$w_m = \left(\frac{3ad}{16}\right)^{1/2} \left(\frac{gC}{\theta_e}\right)^{1/2} z^{1/3}\,, \tag{6.28}$$

$$\theta'_m = (\tfrac{1}{2}d)\, C z^{-1/3}\,. \tag{6.29}$$

A significant feature of the solution is that $R^2 w_m \theta'_m$ tends to zero at the surface; or, although a plume certainly requires some preferred local circumstance to bring it into existence, it will be maintained (following the fluid) without the continuation of any preferred local heat supply. This feature of the solution provides positive reason to expect that the convection will take place in plumes rather than short bursts or bubbles, but it does not give any clue to what the initiating mechanism for an individual plume is most likely to be.

If H_m and $\overset{*}{H}_m$ be used to denote the values of H and $\overset{*}{H}$ corresponding to the velocity and temperature excess at the center of the plume, it follows from the definition of $\overset{*}{H}$ and equations (4.5), (6.28), and (6.29) that

$$\overset{*}{H}_m = \frac{H_m}{\rho\, c_p\,(g/\theta_e)^{1/2} C^{3/2}} = \left(\frac{3\,a d^3}{64}\right)^{1/2}. \qquad (6.30)$$

It was shown by Priestley that, for a field of motion composed of plumes of variable temperature and compensating down-currents of steadier temperature,

$$\overset{*}{H} = \overset{*}{H}_m \beta\ , \qquad (6.31)$$

where

$$\beta = q \frac{\overline{w\,\theta'}}{w_m\,\theta'_m} + \frac{q^2}{p} \frac{\overline{w}\ \overline{\theta'}}{w_m\,\theta'_m}. \qquad (6.32)$$

The bar denotes an average over the cross-section of the individual plume, and p and q are the relative masses of descending and ascending air Evidence was obtained of a remarkable constancy of p and q with height and condition, at values 0.53 and 0.47, respectively, whence β becomes a constant depending only on the similarity functions f and h. As the focus of attention shifts from the individual to the population, it becomes implicit in the whole picture that the process of entrainment must be understood to include the mutual assimilation of the rising elements. With a spreading individual and a dense population, mass continuity can be preserved only in this way.

It has not been taken into account that the lowest layers will generally, in practice, obey a regime of forced convection with $\partial\theta_e/\partial z \propto z^{-1}$, but it is of interest to note that in such an environment the solution for the plume will be of the form $w_m \propto z^{1/2}$, $\theta'_m \propto z^0$, $\overset{*}{H}_m \propto z^{1/2}$, whereas it follows from equation (4.7) that $\overset{*}{H} \propto z^{-1/2}$. The implication is that plumes can and are likely to exist in such a regime, while their relative role in heat flux will be negligible near the surface but will increase with increasing height—an illuminating result in the context of transition of the heat-flux regime discussed in chapter 4.

Another feature of this solution deserving comment is that, following the argument that equations (6.13)–(6.15) should apply in the presence of wind with c the chief wind-dependent parameter, the non-appearance of c in equations (6.30)–(6.32) implies a heat flux independent of wind speed. This completes the outline of a theoretical picture which is quite strikingly consistent with the experimental evidence of chapters 4 and 5. With some wind, however, the adventitious "hot spot" which may be required to bring the plume into existence would be likely to continue as such for some time, whereas the plume is maintained following the wind. Nature would be expected, therefore, to prefer a type of plume which is drawn out into a line in the downwind direction. The two-dimensional plume problem may be followed through by steps analogous to those taken in the axially symmetric case, and it is necessary here to state only the main results. If a', b', c', and d' denote the constants which appear in the corresponding places in the counterparts of equations (6.13)–(6.15), where R is replaced by L with index reduced by 1 in every term, the solution is

$$L = \left(\frac{3a'}{5b' - 6a'}\right) c' z \,, \qquad (6.33)$$

$$w_m = \left(\frac{3a'd'}{5}\right)^{1/2} \left(\frac{gC}{\theta_e}\right)^{1/2} z^{1/3} \,, \qquad (6.34)$$

$$\theta'_m = (d')\, C z^{-1/3} \,, \qquad (6.35)$$

with

$$\overset{*}{H}_m = \left(\frac{3a'd'^3}{5}\right)^{1/2} . \qquad (6.36)$$

The line solution differs from the point solution in its numerical constants but retains the same general properties, and equations (6.31) and (6.32) continue to apply.

We may finally make numerical comparison between the various models, the observed heat flux $\overset{*}{H}$, and typical observed values of

σ_w and σ_T, the root-mean-square fluctuations of w and T at a fixed point. Models A and B will refer to equal-Gaussian profiles for the point and line plumes, respectively. In each case we shall take the observed values of p and q, but β is still not uniquely determinate because the definition of the outer limit of the individual plume remains a matter of convention, as it does in all cognate problems in diffusion; it seems most reasonable to take the effective outer limit at the distance of maximum shear $|\partial w/\partial r|$ or $|\partial w/\partial y|$, since the spreading and dilution of the plume have been attributed to turbulence arising from this shear. The only type of profile that avoids this difficulty is one with an abrupt boundary such as the "top hat," and we include as models C and D the top-hat profiles for

TABLE 6

NUMERICAL VALUES FROM VARIOUS PROFILE MODELS

Model	a,a'	b,b'	d,d'	R/cz, $L/c'z$	$w_m(gC/\theta_e)^{-1/2}z^{-1/3}$	$\theta'_m C^{-1}z^{1/3}$	$\overset{*}{H}_m$	β	$\overset{*}{H}$
A. Point Gaussian.......	2	3	2	1	0.86	1	0.86	0.56	0.48
B. Line Gaussian.......	$\sqrt{2}$	$\sqrt{6}$	$\sqrt{2}$	1.13	1.10	$\sqrt{2}$	1.55	.66	1.02
C. Point top hat	1	2	1	0.43	0.43	$\frac{1}{2}$	0.22	.89	0.20
D. Line top hat	1	2	1	0.75	0.77	1	0.77	0.89	0.69
Observed......	.....	.....	.....	....	$\geqslant 1$	1.2	......	......	0.9

point and line source, with equal spread of momentum and temperature. The numerical values are set out in Table 6. Model C is of academic interest only, and it is not surprising that it leads to unrealistic values. It is significant, however, that the two line-source models give both $\overset{*}{H}$ and θ'_m comparable with the observations, lying, in fact, on each side of the best values determined from experiment, while, as discussed as the end of chapter 5, those for w_m are consistently lower than σ_w, owing to the residual effects of mechanical turbulence, except at the largest $-Ri$, where the observed values approach those predicted by model B. Something considerably better than order-of-magnitude agreement has been obtained, and, when the evidence from the structure records are added, it seems reasonable to claim that the essence of the process has been established; any more stringent comparison between model theory and experiment would be premature until the principles governing the profile functions have been clarified.

7. EVAPORATION

Evaporation and Dew

In the period under review, the work on evaporation, viewed as a problem in turbulent transfer in the atmosphere, to which aspect we are restricted in this volume, has been marked by little influx of basically new ideas, except through the cognate problems in heat and momentum transfer. Consequently, the most practical approaches to the determination of evaporation still lie along two main paths, in each of which some assumption requires to be made relating the transfer mechanism for water vapor to that for either momentum or heat. This is likely to remain the situation until the direct measurement of E by the method of chapter 2 becomes more widely exploited. But so far, as already noted, such measurements of E have tended to lag behind those for H and τ, in part because the measurement of the fine structure of humidity is rather more difficult than that of the other elements, while for the same reason the profile of humidity in the layers very close to the ground has generally received considerably less systematic study than the profiles of wind and temperature. From the findings in chapter 2 that the mechanism of water-vapor transfer tends to follow that of momentum, with some tendency toward that of sensible heat at the larger $|Ri|$, it follows that much that is relevant to the laws of evaporation has, by implication, already been treated in preceding chapters. A few *ad hoc* aspects which are related to the turbulent transfer remain for brief discussion here; and we shall repeat in outline, for reference purposes, the most commonly used approaches and formulas. For a summary of work on evaporation in its wider context, other recent reviews are available (e.g., Deacon, Priestley, and Swinbank, 1958).

The aerodynamic approach involves the assumption that $K_m =$

K_w. From the logarithmic profile (3.1) Thornthwaite and Holzman (1939) derived the expression

$$E = k^2 \rho \frac{(q_1 - q_2)(u_2 - u_1)}{[\ln (z_2/z)]^2}, \tag{7.1}$$

allowing the evaporation to be computed from measurements of humidity and wind at two heights. The formula is valid in adiabatic conditions only; when pressed into service for other occasions, it will underestimate the evaporation in unstable and overestimate in stable conditions. Pending resolution of the full consequences of stability variations, attempts have been made to develop methods of use of equation (7.1) which minimize the possible errors involved. Since the effects of stability are reduced as the surface is approached, Pasquill (1949*a*) advocated that z_1 and z_2 be measured as near to the surface as possible; whence in

$$E = \left\{ \frac{k^2 [1 - (u_1/u_2)]}{[\ln (z_2/z_1)]^2} \right\} \rho u_2 (q_1 - q_2) \tag{7.2}$$

the term in braces is virtually constant for a site of fixed roughness. This procedure involves particularly accurate measurements of the heights z_1 and z_2 and correction for the zero-plane displacement, which in turn requires more accurate and elaborate determination of the wind profile. Furthermore, with a small height interval, the humidity differences are not easy to measure. For a surface as rough as an oat field, these requirements become almost prohibitive except for very elaborate and specialized studies (Rider, 1954*b*), but under the less severe conditions of a clover field the accuracy of the method has been verified by Pasquill.

The difficulties have been relieved to a great extent by Deacon and Swinbank (1956), who dispense with the assumption of logarithmic shape, retaining only that of similarity between the two profiles as embodied in equation (2.6). From equation (3.4), then,

$$E = \rho C_n u_n^2 \frac{q_1 - q_2}{u_2 - u_1}, \tag{7.3}$$

where C_n is the drag coefficient at height z_n and may be obtained from a series of wind-profile measurements in neutral conditions. If z_n can be made sufficiently low to be independent of stability effects, equation (7.3) will then apply under all stabilities. For convenience, z_n and z_1 may be identified and z_2 chosen sufficiently high to allow $q_1 - q_2$ and $u_2 - u_1$ to be measured with accuracy. This form of use does not require the heights themselves to be measured. The danger discussed in chapter 3 remains, however, that fully rough flow may not be maintained at z_n under conditions of the strongest stability, and this must be borne in mind particularly in the problem of estimation of dew. All formulas involving vertical differences of course make stringent demands as to quality of site and exposure.

The second main line of approach has been through the principle of energy balance at the surface and is represented by the equation

$$R_0 = H_0 + G_0 + LE\,, \tag{7.4}$$

where R is net incoming radiation and G is heat flux in the ground, L is the latent heat of evaporation, and the zero suffix denotes surface values. If T and q be measured at two heights, z_1 and z_2, then, assuming that the eddy-transfer mechanisms are identical,

$$\beta = \frac{H_0}{LE} = \frac{c_p}{L}\frac{T_2 - T_1}{q_2 - q_1}, \tag{7.5}$$

and E may be obtained from

$$E = \frac{R_0 - G_0}{L(1+\beta)}. \tag{7.6}$$

This formula has been widely used, and a number of workers have attested that it provides satisfactory estimates of E. This must be true when the ground is thoroughly wet, for then most of the available heat goes in evaporation; with β therefore small, its inaccuracy is of little moment to equation (7.6). In drier conditions β may not be small, and then the dangers are apparent. The only simultaneous measurements of H_0 and E by Swinbank have failed to corroborate equation (7.5), and the use of gradients in equation (7.6) would on these occasions have given erroneously high estimates of evaporation: the same conclusion, with less complete measurements, was reached earlier by Pasquill (1949*b*). Concerning the best choice of z_1 and z_2, the same sort of considerations arise as in the aerodynamic approach.

Eddy-flux measurements of E are not sufficient to allow a profitable analysis of the absolute values of K_w so derived, but studies have been made in which humidity profiles have been carefully recorded in conjunction with evaporation measured from small pots of turf-covered soil in natural surroundings. Pasquill (1949*b*) used a method of analysis closely analogous to equation (3.7). Writing

$$\overset{*}{K}_w = \frac{K_w}{z^2\,(\partial u/\partial z)}$$

here, he found satisfactory agreement between $\overset{*}{K}_w$ and the Holzman and Rossby-Montgomery equations (3.12) and (3.13), in unstable and stable conditions, respectively, with $\sigma_1 = \sigma_2 = 12$. Continuing the investigation and including simultaneous measurements of τ with a drag plate, Rider (1954*a*) found no distinction between $\overset{*}{K}_w$ and $\overset{*}{K}_m$; to the results as already discussed in chapters 2 and 3 may be added that the Rossby-Montgomery relation was found to fit satisfactorily in stable conditions with $\sigma_1 = 8$ or 9.

Equation (3.13) with $\sigma_1 = 10$—roughly the mean of the above values—was used by Monteith (1957) to test against dewfall measured on a balance simulating the natural surface. Good and consistent agreement was obtained through a range of conditions with $Ri_{50\,\text{cm}}$ from $+0.01$ to $+0.2$ and $u_{50\,\text{cm}}$ from 0.7 to 3 m sec^{-1}. The corresponding $K_{50\,\text{cm}}$ lies between 50 and 300 cm^2 sec^{-1}. Angus (1956) reports other measurements of dew which correspond to values of K averaged between surface and screen level from 2.5 cm^2 sec^{-1} at $u_{2\text{m}} = 12$ cm sec^{-1} to 32 cm^2 sec^{-1} at 1.5 m sec^{-1}. These studies emphasize that dewfall is a truly turbulent phenomenon, requiring wind speeds in excess of 50 cm sec^{-1} or so, to give a significant downward transport. The phenomenon is not to be confused with risen dew or distillation from the soil to the grass leaves; within the crop itself, Monteith has inferred values of K_w close to the molecular.

The amount of dew per unit horizontal area, over a large area, is limited by the energy balance. The latent heat released will not exceed the outgoing radiation and must, in general, fall well short of it, heat being brought to the surface from above and below by conduction. The upper limit obtained in this way is of the order of $\frac{1}{2}$–1 mm. per night. Detailed estimates for mid-Europe have been made by Hofmann (1953). However, the strictly one-dimensional pattern may be disturbed through differences in thermal and geometrical characteristics between an individual plant (or artificial structure) and its environment, leading to a process of "dew attraction" which can achieve deposits locally much in excess of the limits defined above. Similar effects on a larger scale become evident as advection between sites of differing characteristics.

Saturated and Open-Water Surfaces

At a saturated ground or free-water surface, $u = 0$ and $q = q_s$, the saturation value. If u and q denote values at a reference height, it is then convenient to define, in close analogy with equation (3.4), an evaporation coefficient,

$$C_E = \frac{E}{\rho\,(q_s - q)\,u}, \tag{7.7}$$

where C_E may be expected to behave in a manner broadly similar to the drag coefficient, C_D. The complete identification $C_E = C_D$ would, however, require not only that $K_W = K_M$ in the fully turbulent region but also that there be similarity in the mechanisms of ultimate transfer of vapor and momentum between the air and its underlying surface.

Over the oceans there is evidence that C_E and C_D are not disparate. Recognizing that G_0 is, on the average, small, Jacobs (1942) employed the energy-balance formulas (7.5) and (7.6), assigning the lower height to the surface and using climatological data to produce most valuable maps of the distribution of E over the major oceans. Independent evaluations were made from equation (7.7), apart from the unknown factor C_E, and a constant value of the latter determined such as to reconcile the total value of E obtained from the two methods. In this way Jacobs obtained

$$C_E = 0.0022\ ,$$

and this is quite consistent with the over-all value of C_D within the accuracy of present knowledge. The result at least suggests that there is no major systematic difference in the transfer mechanisms over the oceans. The transfer of heat is also involved through the use of equation (7.5), but, as has been remarked above, the estimate of E is relatively insensitive to any errors in H_0.

In more detailed aerodynamic formula for use over water, where laminar, transitional, and turbulent layers may exist, the derivation of E has been analogous to that of an electrical current flowing through a number of resistances in series. Considering only a laminar layer of thickness $\delta_1 = 30\nu/u_*$ and a turbulent layer of logarithmic wind profile, Sverdrup (1937) proposed

$$E = \frac{\rho k u_* (q_s - q)}{\ln[(z + z_0)/(\delta_1 + z_0)] + (k\delta_1 u_*)/D}, \qquad (7.8)$$

where D is the coefficient of molecular diffusion for water vapor in air. A value of 0.6 cm. was recommended for z_0, which may be somewhat too high, as more recent evidence, including that of Jacobs, suggests (see also Fig. 4), but the result is not very sensitive to it when a height of several meters is used for z, the main reference level. A number of alternative formulas have been built up on the same lines but involving different assumptions about the details of the various layers. Discussions of their relative merits have been published by Sverdrup (1951) and by Anderson, Anderson, and Marciano (1950).

Reliable experimental data were lacking for some years following the derivation of the various formulas of the type represented by equation (7.8), but recently Marciano and Harbeck (1954) have been able to make discriminating tests against daily evaporation from Lake Hefner. They found that Sverdrup's formula (7.8) gave results in good agreement with observations but that none of those of the same type put forward subsequently could be regarded as satisfactory. Formulas of type (7.1)–(7.3), involving measurements at two heights away from the surface, are, of course, equally applicable to the saturated case, but they involve more exacting instrumental requirements. Of these, the Thornthwaite-Holzman formula alone was tested in detail, with good agreement, on the average, but large individual scatter, due in part no doubt to its known distortion of the true influence of stability.

Good results, comparable with those of equation (7.8), were obtained from the well-known formula of Sutton (1934). Briefly, Sutton considered the incursion of air with constant q_a over a surface with $q_s > q_a$, and his results take into account the progressive change in local rate of evaporation downwind from the leading edge of the wet area. In the two-dimensional problem of an infinite cross-wind strip of length x_0 downwind, with $x = 0$ denoting the upwind edge, the result is

$$\int_0^{x_0} E\,dx = A u^{0.78} x_0^{0.89} \qquad (7.9)$$

in the adiabatic conditions to which the derivation primarily refers. Sutton adapted this solution to give an approximate formula for the total rate of evaporation from a circular area of radius r as proportional to $u^{0.78} r^{1.89}$. Here A and the similar constant in the second expression are proportional to $q_s - q_a$ and contain the molecular viscosity, ν. The application is to smooth surfaces, but Sutton later suggested the replacement of ν by the macroviscosity, $u_* z_0$, for use with rough surfaces, and this was adopted by Marciano and Harbeck in their verification.

The Sutton and Thornthwaite-Holzman forms are both based implicitly on an analogy with the drag force, so that the use of $C_E \approx C_D$ as a working rule gains support from the success of their formulas. The success of two formulas of such contrasting character as Sutton's and Sverdrup's is, however, puzzling and may prove to have been to some extent fortuitous; since one depends on lateral dimensions and the other does not, further tests on water bodies of different sizes should help to clarify the issue. The scale factors in Sutton's formula have not been verified under natural conditions; in fact, a study of available evidence (Kohler, 1954) indicates that evaporation per unit area becomes independent of size for diameters greater than about 12 feet. Sutton's analysis, it should be noted, is non-energetic, and no allowance is made for the progressive adjustment of thermal conditions, and hence of q_s, with distance downwind; such allowance would operate in the sense indicated by Kohler's finding.

The influence of stability also awaits resolution, though it will be less serious than over land and strong evaporation will generally occur only in strong winds where conditions are effectively close to adiabatic. Such effects of stability as remain can be reduced in Sverdrup's formula by lowering the reference height, but the proper method of allowing for them in Sutton's treatment is obscure.

None of the aerodynamic formulas extrapolate suitably into light wind conditions, when evaporation in general will continue so long as external energy is supplied to the system by radiation. On the small scale in the open, measurements by Rohwer (1931) and Penman (1948) have shown that evaporation is proportional to $a + bu$, where a and b are constants. The most recent numerical values assigned by Penman (1956) have been incorporated in a simple type of source-strength formula as

$$E = 0.35\,(e_s - e)\left(0.5 + \frac{u_2}{100}\right), \tag{7.10}$$

with E in millimeters per day, e_s and e the vapor pressures at the surface and screen level in millimeters Hg, and u_2 in miles per day. Penman (1948) combined this with the energy-balance approach in such a way that conditions at the surface could be eliminated from the equations, thereby providing a method of calculating 24-hour evaporation from observations available in the normal climatic records. The same was done independently by Ferguson (1952), along similar lines but with fewer approximations.

Penman also measured daily evaporation from bare soil and short-grass surfaces, in containers similar to his tank and similarly exposed, flush with the surroundings, and adequately supplied with water. He found that the bare-soil evaporation correlated fairly well with, and averaged 90 per cent of, that from the water surface. There was a tendency toward a maximum ratio in the spring, the presence of seasonal effects suggesting that thermal inertia and advective influences cannot be neglected. For the short-grass surface the ratio to the water ranged from 0.6 in midwinter to 0.8 in midsummer, with an average of 0.75. The data relate to southern England. In interpretation it must be borne in mind that evaporation from the water may continue through the night, while that from the grass is inhibited by stomatal closure.

8. EVOLUTIONARY ASPECTS OF ENERGY TRANSFER

The General Problem

The previous chapters have dealt with transfer processes and their relationship to the physical condition of the surface layers as these exist at any one place and time. In the wider field of synoptic meteorology the interest must turn toward the pattern in space and time of the surface heating or, more generally, that of the total energy source. In particular, the conformation or non-conformation of this pattern to the scale of the typical weather systems must be a significant factor in the development of circulations on this scale.

Two basic situations may be recognized in which a semi-infinite column of the atmosphere may gain or lose heat by conduction. The first will arise when an air mass comes into contact with an underlying surface with which it is not in thermal adjustment; there will then be a transfer of heat from one medium to the other at a rate governed by the difference in temperature between them and by the wind speed. The second circumstance is when heat is supplied at, or removed from, the interface by an outside agency (radiation); the gain or loss will be shared between the two media in proportions determined by their physical characteristics, but this is not to be regarded as a transfer of heat from one to the other, as in the first situation. A formal separation of the two processes (Priestley, 1957) shows that they can, with fair approximation, be treated independently and that generally in nature one or the other will be predominant on a given occasion. The inclusion of evaporation complicates the picture, but it is a complication which affects the detail rather than the general character of the results, and it will bring into clearer relief the true nature of the factors which control the evolution to ignore evaporation in the first instance. The problem of evolution of heat flux into the atmosphere may then be defined as either (*a*) the sharing of R_0, the net incoming radiation, between H_0 and G_0, or (*b*) the determination of H_0 ($= -G_0$) as a function of time (or distance) when an air mass invades a surface to which it is not thermally adjusted.

The governing equation of vertical diffusion is generally taken as

$$\frac{dT}{dt} = \frac{\partial}{\partial z}\left\{K\left(\frac{\partial T}{\partial z} + \Gamma\right)\right\}, \tag{8.1}$$

which, when conditions are sufficiently near to horizontal uniformity, becomes

$$\frac{\partial T}{\partial t} = \frac{\partial}{\partial z}\left\{K\left(\frac{\partial T}{\partial z} + \Gamma\right)\right\}. \tag{8.2}$$

Sutton (1953) has given a full discussion of a number of solutions of equation (8.2) by different workers, which we shall not duplicate here: the solutions may be classified into those which relate to the propagation of the diurnal temperature wave through the atmosphere and those which treat the adjustment process with a boundary condition prescribing either T_0 or H_0 as a generally constant function of time. A contribution of subsequent work has been to emphasize more forcibly that both problems are essentially problems of conduction in two media. Primed symbols will be used throughout this chapter to distinguish the underlying medium. The equation for T' is

$$\frac{\partial T'}{\partial t} = \frac{\partial}{\partial z'}\left(\kappa\frac{\partial T'}{\partial z'}\right), \tag{8.3}$$

where z' denotes depth and κ will have a molecular or an eddy connotation for land (or ice) and sea regions, respectively. Then T_0, H_0, and G_0 generally emerge as dependent quantities of the solution. It will be shown that in special circumstances it is valid to prescribe the form of T_0, but the assignment in advance of H_0 or G_0 is not really appropriate to naturally occurring processes; both H_0 and G_0 in fact are found to be very sensitive to the physical properties of the underlying medium.

To the extent that a medium may be characterized by a single value of K or κ, certain general results of a comparative nature follow at once. The substitution $z = \zeta\sqrt{K}$ or $z' = \zeta'\sqrt{\kappa}$ reduces all media to a common basis, so that in any application the depth of penetration is proportional to $\sqrt{K}$ or $\sqrt{\kappa}$. The heat capacity per unit volume is proportional to ρc (ρc_p in the atmosphere), and the total heat capacity of the depth penetrated is accordingly proportional to $\rho c\sqrt{\kappa}$: that is to say, the rate of heat flux consequent on an imposed temperature regime at the boundary will be directly proportional to $\rho c\sqrt{\kappa}$, or the rate of temperature rise for an assigned boundary heat flux will be inversely proportional to it. These relationships are independent of the time pattern of the imposed boundary heating (or temperature) regime. It appears, then, that the most directly significant physical properties of the medium are the thermal diffusivity κ and the combination $\rho c\sqrt{\kappa}$, and this latter, in virtue of its significance as well as for convenience, merits a distinguishing name. It will here be referred to as the *conductive capacity* of the medium.

Typical values of the relevant physical properties of different media are set out in Table 7. The values given for stirred water and stirred air are, of course, in round numbers only and in the latter case apply to the air at some distance from the surface. It may be noted that most wet soils have about the same conductive capacity as still water; they will be treated as a single group in subsequent calculations. Dry conditions will commonly obtain, if at all, only through a rather limited depth close to the surface; this is significant in applications involving the longer periods of time for which, as will become clear in what follows, the main control is exercised by conditions at greater depth. Similar considerations apply to snow-covered ground.

On the basis of constant κ (and K) it is clear from the equations that the depth affected, after conduction has been in operation for time t, is proportional to $(\kappa t)^{1/2}$; correspondingly, the amplitude of the diurnal or annual temperature wave of period $2\pi/\nu$ decreases by a factor $1/e$ for each depth interval $(2\kappa/\nu)^{1/2}$ or height interval $(2K/\nu)^{1/2}$. The values of these "damping depths" follow from Table 7.

There is a strong differentiation with respect to thermal diffusivity, and it is seen at once that any heat sharing or heat transfer between the atmosphere and its underlying medium will involve a relatively small depth of soil, snow, or ice, a moderate depth of ocean, and a relatively large depth of atmosphere. With regard to conductive capacity, the relative values will depend on stability and hence on the direction in which heat is being transferred. Generally, that of the atmosphere will be far smaller than that of the ocean; under conditions of heating, that of the atmosphere will compare with that of most soils and be distinctly larger than that of snow or dry sand. The consequences of contrasts in conductive capacity will depend on which type of situation, *a* or *b* in Table 7, obtains.

The Diurnal Variation

Lettau (1951) has made a detailed analysis of the daily variation in two media under the assumption that K follows the forced-convection relation,

$$K = ku_* \,(z + z_0)\,, \tag{8.4}$$

TABLE 7

Physical Properties of Different Media

Medium	ρc (Cal/Cm^3 °C.)	κ or K (Cm^2/Sec)	$\rho c\sqrt{\kappa}$ or $\rho c\sqrt{K}$ (Cal, Cgs Units)	Authority
New snow	0.03	0.006	0.002	Geiger (1948)
Old snow	0.22	0.003	0.012	Geiger (1948)
Dry sand	0.3	0.0013	0.011	Geiger (1948)
Wet sand	0.4	0.01	0.04	Geiger (1948)
Sandy clay (15 per cent moisture)	0.6	0.0037	0.037	Lettau (1951)
Organic soil	0.57	0.005	0.04	Geiger (1948)
Wet marshy soil	0.7	0.003	0.038	Geiger (1948)
Ice	0.45	0.012	0.05	Geiger (1948)
Still water	1.0	0.0015	0.039	Physical tables
Stirred water				
a) Great stability, moderate currents	1	0.1	0.32	Sverdrup *et al.* (1942)
b) Moderately stable, strong currents	1	50	7	Sverdrup *et al.* (1942)
c) Homogeneous, strong currents	1	300	17	Sverdrup *et al.* (1942)
Still air	3×10^{-4}	0.2	1.3×10^{-4}	Physical tables
Stirred air				
Very stable	3×10^{-4}	10^3	0.01	Various
Neutral	3×10^{-4}	10^5	0.1	Various
Very unstable	3×10^{-4}	10^7	1	Various

while κ is constant. The boundary conditions are that $T_0 = T_0'$ and $R_0 = LE_0 + H_0 + G_0$, while both H and G tend to zero at large height and depth. For simplicity, E_0 was taken to be constant, and a simple harmonic form was assumed for R_0,

$$R_0 = R_{00} + r_0 e^{-i\nu t} ,$$

where R_{00}, r_0, and the corresponding terms in H, G, and T below are all constant.

The mathematical analysis is intricate, but the main results can be summarized in terms of the amplitude and phase of H_0, G_0, and T_0 relative to those of R_0. Writing

$$H_0 = H_{00} + h_0 e^{-i(\nu t+\alpha)} ,$$

$$G_0 = G_{00} + g_0 e^{-i(\nu t+\beta)} ,$$

$$T_0 = T_{00} + \Delta T_0 e^{-i(\nu t+\gamma)} .$$

Lettau finds the relative amplitudes of G_0 and H_0 to be given by

$$r_0 = h_0 \cos \alpha + g_0 \cos \beta ,$$

together with

$$\frac{g_0}{h_0} = \frac{\pi c' \rho'}{k u_* c_p \rho} \{\nu\kappa\,(j^2 + \tfrac{1}{4})\}^{1/2} , \tag{8.5}$$

where

$$j = -\frac{2}{\pi} \log_e \left\{1.781 \left(\frac{\nu z_0}{k u_*}\right)^{1/2}\right\} .$$

Values of j commonly lie between 3 and 6, so that for practical purposes the term in braces in equation (8.5) may be written $j\sqrt{\nu\kappa}$. The amplitude of surface temperature and the phases α, β, and γ are given by the equations

$$\Delta T_0 = \frac{g_0}{c' \rho' \sqrt{\nu\kappa}} , \tag{8.6}$$

$$\alpha = \gamma + \tan^{-1} \frac{1}{2j} = \gamma + \phi , \tag{8.7}$$

$$\beta = \gamma + \frac{\pi}{4} , \tag{8.8}$$

$$\tan \gamma = -\frac{g_0 + h_0 \sqrt{2} \sin \phi}{g_0 + h_0 \sqrt{2} \cos \varphi} , \tag{8.9}$$

which complete the solution. A typical set of results for the diurnal cycle have been calculated by Lettau and are reproduced here in Table 8.

This is the only analysis of the two-medium problem that includes explicit allowance for the influence of wind and roughness. As would be expected on intuitive grounds, the relationships are rather insensitive to surface roughness per se, an increase in z_0 by a factor of 10^4 approximately halving the value of j over the relevant range of u_*; and u_* itself affects the results mainly through the factor outside the braces in equation (8.5). The analysis must clearly become invalid when the wind is light, for then K will not become small as in equation (8.4) during the daylight hours but will tend to take up the value appropriate to free convection.

TABLE 8

TYPICAL CHARACTERISTICS OF THE DIURNAL WAVE AFTER LETTAU
($u_* = 20$ Cm Sec^{-1}; $z_0 = 0.1$ Cm)

Ground Type	Conductive Capacity ($\rho c\sqrt{\kappa}$)	Amplitude Ratio (g_0/h_0)	Lag of Maximum after Maximum of R_0 (Hr.:Min.)		
			T_0	G_0	H_0
Concrete	0.057	2.56	2:18	−0:42	1:53
Sandy clay	0.036	1.65	2:02	−0:58	1:37
Snow	0.005	0.24	0:54	−2:06	0:29
Ocean					
$\kappa=1$	1	46	2:58	−0:02	2:33
$\kappa=100$	10	460	2:59	−0:01	2:34

Unfortunately, the amplitudes g_0 and h_0 provide no information as to the amounts of heat gained by atmosphere and ground in a complete cycle. To obtain these for the diurnal cycle, Lettau carries through the parallel calculations of the annual cycle, so obtaining H_{00} and G_{00} at any phase of the year. However, the physical relevance of treating the annual wave as a phenomenon of small-scale diffusion cannot be seriously upheld, nor would equation (8.4) be expected to be appropriate to any such context. In any event, it would seem that the analysis of the diurnal cycle could properly be made almost, though admittedly not quite, self-sufficient; that is to say, the sharing of the heat made available in the 24 hours is determined largely by the properties of the two media and is influenced to only a minor extent by conditions which are a residual from previous seasons.

The main shortcoming of equation (8.4) in the present context, as in the problem of approach to the geostrophic wind discussed earlier, is that it does not recognize the bounded form of K at greater heights, and this has repercussions on the deductions made about happenings at lower levels. The issues involved here do not appear to have been discussed at any length in the textbooks; but they deserve our detailed attention because they relate to the possibility of making effective application of the simpler $K(z)$ forms whose parameters we might reasonably hope to estimate from conventional synoptic data backed by experience. The simpler algebraic K forms must all represent distortions in some layer or layers, and the question arises as to where the sacrifice of accuracy in $K(z)$ will have the least serious effect on the magnitude and pattern of evolution of H_0 which are deduced.

In general, the positive values of R_0 and H_0 during the midday hours considerably exceed their negative values at night, so that happenings during the midday regime will play the major part in determining the net gain of heat to the atmosphere over a day or more. Sutton (1948) has shown that on fine days nearly all the H_0 around the maximum epoch penetrates to heights of 100 meters and beyond or, as is seen by reference to Table 2, to levels where equation (8.4) has ceased to apply and a regime of more nearly constant K is established. This means that an important measure of control of the ability of the atmosphere to acquire heat is exercised through conduction in the region of constant K and that an unbounded form of K such as (8.4) must produce distortions not only in the constant K levels but also through the main depth below.

Ideally, a form for K should be used which recognizes the bounded condition above 50 or 100 meters, the possibility of laminar flow very close to the surface, and, in the daytime at least, two other regimes (forced and free convection) in different layers between. Accepting the need for simplification, however, recent papers have considered the effect of a periodic linear variation (Poppendiek, 1952) and of forms which increase smoothly (Staley, 1956) or otherwise (De Vries, 1957) to an upper bound. The general implication of these formal studies is that an unbounded K introduces distortion at all levels and, in consequence, erroneous values of the heat flux. A number of different bounded forms, on the other hand, give results which converge at the greater heights when the bounding values are the same and may therefore be expected to represent more or less correctly the heat transferred to these regions over a full cycle, even though some discrepancies very close to the surface

TABLE 9

COMPARISON OF DIURNAL WAVE WITH TWO MODELS

	HEIGHT (METERS)						
	2	5	10	20	50	100	200
K (10^5 cm^2 sec^{-1})							
a	1.2	1.2	1.2	1.2	1.2	1.2	1.2
b.............	0.024	0.06	0.12	0.24	0.6	1.2	1.2
Amplitude of T (° C.)							
a.............	2.3	2.3	2.3	2.3	2.1	1.95	1.65
b.............	3.5	3.2	2.9	2.6	2.2	1.95	1.65
Amplitude of H (10^{-3} cal cm^{-2} sec^{-1})							
a.............	1.99	1.98	1.97	1.93	1.83	1.68	1.41
b.............	1.99	1.98	1.97	1.93	1.83	1.68	1.41
Phase of T (radians)							
a.............	0.79	0.79	0.80	0.82	0.87	0.96	1.13
b.............	0.50	0.57	0.64	0.73	0.87	0.99	1.16

will appear. As an example, we have given in Table 9 a comparison of results obtained from two models, one (*a*) with $K = 1.2 \times 10^5$ cm^2 sec^{-1} at all levels and the other (*b*) where K increases linearly to this value at 100 meters and remains constant above; *a* is given by the familiar classical solution, and the calculations for model *b* were carried through by De Vries (1957), the figures quoted here being kindly supplied by him. The boundary condition is that of a periodic heat flux of amplitude 2×10^{-3} cal cm^{-2} sec^{-1} and phase zero (positive phases in Table 9 represent later occurrences of maxima, etc.). There is no important difference between the models in the amount of heat penetrating to any level. It appears that in applications where the phase of T close to the surface is important—for example, the timing of frost danger—the variation of K with height near the surface is significant but that in applications concerned with the magnitude of the heat flux, the variation of K with height can be neglected without serious error.

With regard to other than linear forms for the region where K varies strongly with height, it must be recalled from the discussion in chapter 4 that these generally represent reflex adjustments to the heat-flux regime. Treatments of the diurnal problem which take such forms as predetermined are in danger of misrepresenting the true chain of cause and effect.

We may conclude that the main control of the sharing between the two media of the heat made available over any considerable length of time is exercised by the values of K and κ at some distance from the surface, where these are more nearly constant. There is much merit, therefore, in the classic treatment of the problem through the Fickian, or constant K, equations of diffusion, as first exploited by Schmidt (1918). He solved the two-medium problem in the case of a simple harmonic source of heat at the boundary, but the solution may be obtained without any restriction on the form of the periodic variation. With T' in equation (8.3) the counterpart of $T + \Gamma z$ in equation (8.2), the conditions in the two media are the same except insofar as K differs from κ, and the substitution $z = \zeta\sqrt{K}$ and $z' = \zeta'\sqrt{\kappa}$ makes the symmetry complete. It follows, then, that, at all times and irrespective of the form of the time variation imposed at the boundary,

$$\frac{H_0}{G_0} = \frac{\rho c_p K\,[(\partial T/\partial z) + \Gamma]_0}{\rho' c' \kappa\,(\partial T'/\partial z')_0} = \frac{\rho c_p \sqrt{K}\,[(\partial T/\partial \zeta) + \Gamma\sqrt{K}]_0}{\rho' c' \sqrt{\kappa}\,(\partial T'/\partial \zeta')_0} = \frac{\rho c_p \sqrt{K}}{\rho' c' \sqrt{\kappa}}. \qquad (8.10)$$

Rates of change of surface temperature are proportional to R_0 and inversely to the sum of the conductive capacities.

The application of these rules affords a ready explanation for well-known comparative features of climate, such as the strong contrasts in diurnal range of temperature in the air over sea, over different types of soil, and over snow fields. Quantitatively there is value, in that, in conjunction with data such as that of Table 7, a means of estimation is provided in simple and synoptically appreciable terms. Practical methods for estimating the total energy available for sharing, or 24-hour integration of R_0, have been given by Korb, Michalowsky, and Möller (1956); this energy may amount to as much as 350 cal cm^{-2} per day in spring in latitude 30°, and presumably more in summer and in lower latitudes.

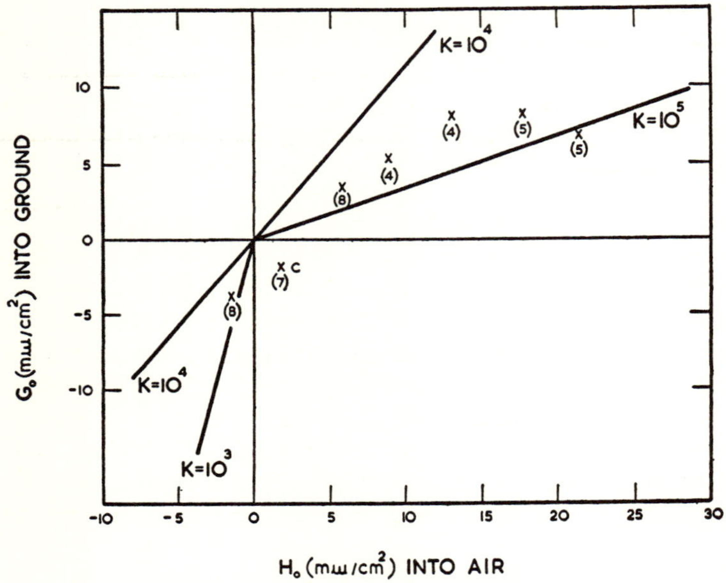

FIG. 21.—Simultaneous heat flux into ground and air

No searching test of the accuracy of the sharing formula (8.10) is possible at present. Only Swinbank (1955) has made simultaneous measurements of H_0 and G_0, and these are shown in Figure 21, the individual scatter being reduced by using only the H_0 values at low levels and plotting average G_0 in ranges of H_0. The sloping lines are drawn from equation (8.10) for various values of K, the value of $\rho' c' \sqrt{\kappa}$ having been determined as 0.034 under the conditions of the experiments. The number of individual pairs of measurements are indicated in brackets. This preliminary analysis suggests that a useful estimate of the sharing over land may be obtained by taking

$K = 10^5$ cm^2 sec^{-1} by day and 10^3 at night. Owing to the phase distortion in the Fickian analysis, equation (8.10) will be quite inapplicable at transition times, as evidenced by the point marked C in Figure 21, but this should not cause serious error in apportioning the 24-hour totals for the heat received by ground and air, respectively.

The inverse problem to that discussed above arises where observed amplitudes and phases of the diurnal wave at different heights are used to calculate values of K. Rather little work has been done on this since the stage summarized by Sutton was reached. The classical calculations were by Best (1935), who allowed K to increase with height stepwise and, testing the results from discrete intervals against the continuous variation given by

$$K_h \propto z^p , \tag{8.11}$$

found $p = 1.8$ to be appropriate to days of strong heating. The procedure has been criticized, however, because continuity of heat flux is not preserved at the steps, and a reanalysis of the same data along more rigorous lines by De Vries (1956) leads to a p between 1 and $\frac{4}{3}$. This result is certainly of interest in the light of the discussion of chapter 4, but De Vries (1957) shows that values of K obtained in this way are so sensitive to apparently insignificant changes in observed amplitude and phase that no great weight can be attached to them and preference must be given to more direct determinations.

Air-Mass Adjustment

When an air mass comes into contact with a surface to which it is not thermally adjusted, an actual transfer of heat between the media results, and a new process of air-mass modification is set in operation. This must occur to a greater or less extent following every frontal passage.

When this takes place over water, it follows from a comparison of the conductive capacities (Table 7) that the temperature change in the water will be quite negligible in relation to that of the air, so that the problem can be treated as that of flow, with the boundary at a prescribed surface temperature. This was the problem solved by G. I. Taylor with constant K, which provided early knowledge of the magnitudes of K appropriate to the atmosphere. Taking the initial temperature of the air as $T_i - \beta z$ and that of the water as T'_i, where T_i and T'_i are constants, the solution is

$$T = T_i - \beta z + (T'_i - T_i)(1 - \text{erf } \xi) , \qquad (8.12)$$

where $\xi = z/\sqrt{4Kt}$. Taylor found $K \approx 10^3$ cm^2 sec^{-1} for winds of force 2, and 3×10^3 with force 3, as appropriate to stable conditions in the lowest 1,000 feet. In equation (8.12) and the ensuing formulations, e.g., equations (8.17)–(8.19), the solutions for the air temperature are expressed as time variations referred to axes moving with the wind. The wind speed is assumed constant with height, and horizontal diffusion is neglected.

Here again the validity of the Fickian (constant K) approach will depend on the distance and time scale of the process under consideration. Taylor's distance scale was of several hundred miles. Successive profiles up to 30 miles from shore as warm air moved seaward over Massachusetts Bay were measured by Craig (1949) and could not be reconciled with a model either with K constant or with K increasing linearly throughout the layer. A linear increase to a value of $1\text{–}2 \times 10^3$ at 1–2 meters height, remaining constant at higher levels, provided the best fit in the lighter wind conditions. However, with winds of 20 mi hr^{-1} and above and consequently reduced stability, K_w was found to attain values of 5×10^4 around 200–250 feet, with a rather rapid decline above that level. It is not without significance that this maximum became progressively less pronounced with increasing distance from shore. Craig's work provides an illustration of the increasing appropriateness of the Fickian form as the scale of the application increases.

Quite a different approach was employed by Burke (1945) in the case of polar continental air moving over warmer sea. By analogy with equation (3.4), written as

$$\tau = \rho C_D u^2 ,$$

Burke assumed that the transfer of sensible heat from the sea into the air would be given, in our present notation, by

$$H_0 = -G_0 = c_p \rho C_D u (T'_0 - T_0) \qquad (8.13)$$

and the evaporation by

$$E_0 = \rho C_D u (q'_0 - q_0) . \qquad (8.14)$$

It is clear that the symbols inside the parentheses no longer denote truly surface values but refer to levels a short distance from the surface, where "representative" conditions may be said to exist. Implicitly, here, a forced-convection type of relation has been averaged through the layer between these levels, the value of K and its variation with height being absorbed in a single transfer coefficient. We shall adopt, with u referred to heights of about 10 meters,

$$c_p \rho C_D = A = 7.8 \times 10^{-7} \text{ cal cm}^{-3} \text{ °C.}^{-1}, \qquad (8.15)$$

which corresponds to $C_D = 0.0025$, appropriate to the stronger wind speeds (see Fig. 4, p. 22). This was the value used by Burke in his calculations. Burke did not apply the diffusion equation but assumed that the modified air would lie wholly below the unmodified air and that its profile would follow the dry adiabatic to the condensation level and the saturated adiabatic above; the positions of the adiabatics were determined by equating the total heating and humidity increase over the period in question to the time integrals of equations (8.13) and (8.14). The results were embodied in a set of diagrams which allowed the solution to be obtained graphically, given the initial lapse rate, wind speed, and temperature and humidity differences. In this way, good agreement was obtained between the values of T_0 observed and calculated after various distances of trajectory off the east coast of North America.

Burke's method is the simplest for application and is sound, provided that there is a pronounced initial stability to confine the depth affected by the modification process; but in genuine polar outbreaks, with long trajectories, this is not the case, and the method breaks down. The sensible heat flux alone may then exceed a sustained rate of 50 cal cm^{-2} hr^{-1} (Craddock, 1951), and the total rate of energy gain to the atmosphere in these conditions will exceed that achieved by any other process.

The diffusivity of the atmosphere at large may be allowed for by reintroducing equation (8.1), K now applying to the region above the forced-convection boundary layer, which latter has been incorporated in equation (8.13). Reducing the surface temperature of the water, T'_0, to zero for convenience, the boundary condition now is

$$H_0 = -\rho c_p K \left(\frac{\partial T}{\partial z} + \Gamma\right)_0 = -A u T_0 ; \qquad (8.16)$$

and if $T_i - \Gamma z$ is the initial temperature distribution in the air, then the solution of the Fickian equation is given (Priestley, 1957) by

$$\frac{T+\Gamma z}{T_i} = \psi(\xi, \eta) = \text{erf}\ \xi + e^{2\xi\eta+\eta^2}\{1-\text{erf}(\xi+\eta)\}, \quad (8.17)$$

where $\eta = (Au/\rho c_p)(t/K)^{1/2} = C_D u(t/K)^{1/2}$. When the initial lapse rate differs from Γ by a constant amount γ, i.e., initially $T = T_i - (\Gamma - \gamma)z$, the solution becomes

$$T+\Gamma z = \left(T_i - \frac{K\gamma}{C_D u}\right)\psi(\xi, \eta) + \gamma z + \frac{K\gamma}{C_D u}. \quad (8.18)$$

We shall be concerned here with the simpler form (8.17). Putting $\xi = 0$ and using equation (8.16), the evolution of surface temperature and heat flux are found to be

$$\frac{T_0}{T_i} = \frac{H_0}{H_{0i}} = e^{\eta^2}(1 - \text{erf}\ \eta), \quad (8.19)$$

where i is used to denote initial values. This equation establishes the time scale of the process, a given fraction of the initial value being obtained after a time of travel which is proportional to K but inversely proportional to the square of u and of C_D or A.

The use of the bulk transfer type of boundary condition (8.13) in conjunction with the Fickian equation provides a formulation which is entirely in terms of bulk properties and may on this account be held to be more closely aligned with the type of information available directly or by inference in the synoptic framework. Inevitably, as in the diurnal problem, some of the fine detail of what is happening very close to the surface is lost. Here again, on any substantial time scale, the main control must reside in the constant K regions because of the relatively small heat capacity of the layers where K varies strongly.

Having simplified the formulation by this means, we are able to make some progress with the parallel problem of invasion over land, which has hitherto received little attention. Reference to the conductive capacities in Table 7 shows that modification of the temperature of the underlying medium will then proceed as rapidly as that of the air in general and rather more rapidly in the case of a cold outbreak, when the air may be taken to be in a neutral or slightly unstable condition. Treatment of T'_0 as a constant, or any assignment of its form in advance, is therefore inadmissible.

As a first approximation to the case of heating over land, then, we may reverse the roles of air and underlying medium in the previous solution, obtaining the evolution of $H_0 = -G_0$ on the assumption that the actual temperatures in the air will change rather little over a period in which those in the ground will change substantially. The previous solution then applies, with, of course, $\Gamma = 0$ and obvious changes in connotation of the other variables. It may be shown from equation (8.19) that, for large η,

$$\int_0^t H_0 dt \sim \frac{2}{\sqrt{\pi}} T_i \rho c \sqrt{\kappa t} . \tag{8.20}$$

This is a practical result of considerable significance. For media of low conductive capacity, η will relatively rapidly approach the values at which the asymptotic expression (8.20) applies. The total heat transferred over such a period is seen to be proportional to the conductive capacity and the initial temperature difference but to be independent of the wind speed and the magnitude of the transfer coefficient, A. Thus, provided that we are concerned with periods of sufficient length, the uncertainty which attaches to the most appropriate value of A for a land surface ceases to be of practical concern.

In order to illustrate the typical time factors involved, calculations have been made from equation (8.19) on the basis of a wind speed of 10 m sec^{-1} and an initial temperature difference of 10° C., with the value of A quoted above. The results are set out in Table 10. The lower rows refer to air in contact with water, which may be taken without significant error to be of infinite conductive capacity; the upper rows refer to various solid underlying media in contact with air, which is taken to be of infinite conductive capacity as a first approximation for neutral or unstable air. With stable air in contact with a solid surface, it is clear from the other figures that the time scale will be very short indeed and that no significant amount of heat will be transferred. A final column, included for comparison, relates either to a wind of 20 m sec^{-1} with the same A or to a wind of 10 m sec^{-1} with $A = 15.6 \times 10^{-7}$. The range of applicability of equation (8.20) is illustrated by this and other columns in the table.

Over land surfaces in general, therefore, H_0 will decline by an order of magnitude within about 24 hours at most, while the total amounts of heat transferred will generally be smaller than those obtainable by the air as a share of the 24-hour insolation, as indicated by equation (8.10). Over sea the converse of the last statement will

TABLE 10

EVOLUTION OF HEAT FLUX WITH DIFFERENT MEDIA
(Wind Speed 10 M Sec^{-1}; Initial Temperature Difference 10° C.)

Medium	Conductive Capacity	Time for H_0 to Fall to Fraction of Initial Value			Heat Transfer (Cal Cm^{-2})			
		$\frac{1}{2}$	$\frac{1}{5}$	$\frac{1}{10}$	First 12 Hours	First 24 Hours	Second 24 Hours	First 24 Hours ($u=20$ M Sec^{-1})
New snow...	0.002	4 sec.	45 sec.	3 min.	5	7	3	7
Old snow, dry sand......	0.01	100 sec.	19 min.	80 min.	22	32	14	33
Wet soils....	0.04	26 min.	5 hr.	22 hr.	75	110	55	120
Ice..........	0.05	40 min.	8 hr.	35 hr.	90	140	70	150
Air ($K=10^3$)...	0.01	100 sec.	19 min.	80 min.	22	32	14	33
($K=10^5$)...	0.1	$2\frac{1}{2}$ hr.	30 hr.	6 days	145	235	130	275
($K=10^7$)*..	1	11 days	130 days	570 days	310	580	500	1,000

* With $K = 10^7$, some correction would be due for the effects of the "ceiling" because the modification (eq. [8.17]) will reach to the tropopause within 12 hours. With $K = 10^5$, however, temperature change will be confined below 3 km. for the first 48 hours.

be true. In both instances the splitting of the modification process into two distinct problems has been largely justified by the numerical magnitudes which emerge in the solutions.

Where the conductive capacities of the two media in contact are comparable, an accurate solution is possible only by treating the full two-medium problem, and the relative motion of the two media presents complications which stand in the way of an analytical solution. At a time, say t hours, behind a fresh outbreak, the ground will have experienced modification by contact with fresh air for a period of t hours; on the other hand, the time of contact which the air at this point and this stage will have experienced will be related not to t but to the length of the trajectory over representative terrain. There exists, therefore, no unique relation between the one time and the other, and numerical *ad hoc* solutions are the best that can be expected. A method of fair accuracy has been described by Priestley (1957), and the calculations have been carried out for various underlying surfaces with the same initial conditions as in Table 10. Figure 22 shows the total heat, $\int H_0 dt$, taken up by the air at different distances behind the front in the first 24 hours following the outbreak. The calculations were made with $K = 10^5$ and

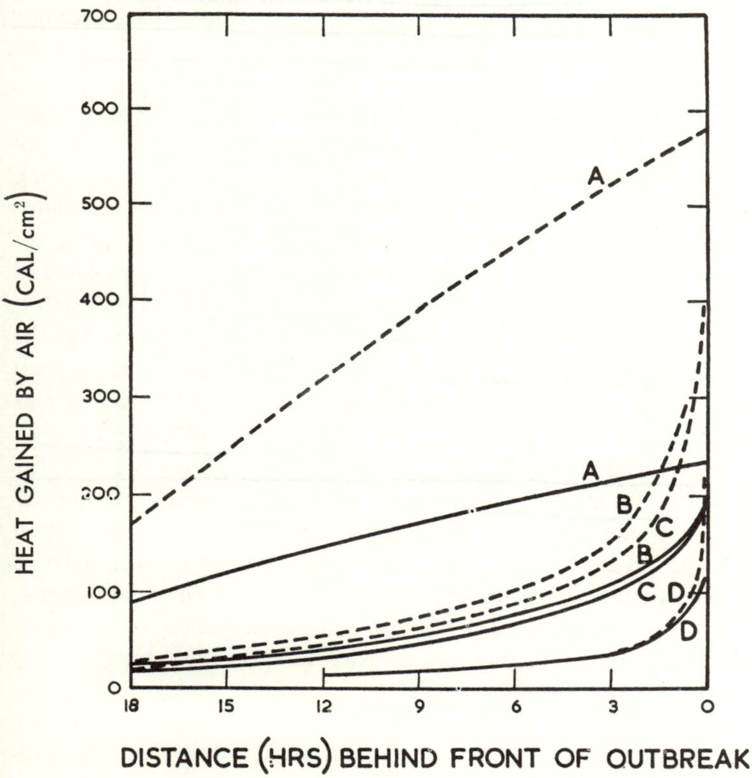

FIG. 22.—Heat gained by air in 24 hours in a cold outbreak over different underlying surfaces: *A*, water; *B*, ice; *C*, wet soils; *D*, dry sand. Full and broken curves denote $K = 10^5$ and 10^7 cm² sec⁻¹, respectively. $T_i = 10°$ C., $u = 10$ m sec⁻¹, $A = 7.8 \times 10^{-7}$ cal cm⁻³ ° C.⁻¹.

10^7 cm^2 sec^{-1} and four typical underlying surfaces. As has been implied, with the same initial conditions (which include the same coefficient of heat transfer and hence the same initial rate H_{0i}) very different total heat transfer will result according to the type of underlying surface. Each set of four curves converges to the same ordinate at the right-hand border of the diagram, since just behind the front the air is all the time in contact with unmodified ground and the value here is determined by K and is independent of κ. The diagram illustrates, among other features of interest, that over land the temperature response of the soil becomes the dominating factor, and a highly unstable air mass gains little more heat than one which is only moderately unstable, except immediately behind the front.

The invasion problem as a whole offers scope for more detailed analysis, for which numerical methods would appear the most appropriate, but the approximate treatment above has served to bring out certain new and generally significant features of the solutions. The main practical result is to provide the time or distance factor involved in the process of adjustment to an air-mass invasion and to show how strongly this is at the mercy of the physical characteristics of the underlying medium. Second, it has been shown that the roughness of natural terrains, which can be held to be characterized by the value of A (and may be distinguished from site roughness, as commonly indicated by the parameter z_0), has little effect on the heat transfer in applications concerned with a time scale of the order of 24 hours or longer. Finally, it may be observed in Figure 22 that not only the degree of heating but also the pattern of heating will be different as between a land and a sea trajectory, and this may have important consequences in cyclonic development because, according to Petterssen (1956, chap. 16), here it is the second space derivative of the heating function that is most significant.

Latent Heat Effects

We conclude by discussing briefly the manner in which the foregoing results will be influenced by latent heat effects.

It has been shown that the exchange of heat between air and ground cannot normally continue for any length of time, owing largely to the fact that the ground is a poor conductor, so that its near-surface temperature moves rather rapidly toward that of the air. An exception arises when a wet soil (or snow or ice surface) is at 0° C.; the sensible heat transferred to or from the atmosphere can then be drawn from, or taken up by, processes of freezing or melting. No strain will then be placed on the conductive capacity of the ground, and a high value of H_0 can be maintained for a much longer time. This appears to explain the high probability of occurrence of fog at temperatures close to 0° C. (Petterssen, 1939), since persistence of fog requires continuance of the heat withdrawal at a high rate in order that the rate of formation of droplets may balance the gravitational settling.

When air invades a wet surface to which it is not physically adjusted, the evaporation will be subject to a pattern of evolutionary control similar to that of the heat transfer. In addition, it must be recognized that the heat used in evaporation must be drawn from one or another of the media and so may react on the temperature response and hence on the heat exchange. With cold air invading a warm surface, evaporation will occur as the air warms up, but the underlying medium, being the warmer, must provide the energy for this evaporation and will therefore cool the more rapidly. In the ocean this is of no consequence, for the conductive capacity is effectively infinite and we may regard the evolution of heat exchange and evaporation as operating independently and the energies so gained by the atmosphere as additive. Over land, however, the higher initial rate of energy loss will affect the temperature response of the ground in the same way as a higher value of A or C_D in equation (8.13); and we would expect that, for periods long enough for the asymptotic relations such as (8.20) to apply, the increase in atmospheric energy represented by its latent component will be gained only at the expense of a corresponding reduction in the sensible heat gain.

When warm air invades a cold surface, the transfer of moisture may be in either direction according to the humidity. If the air is sufficiently moist, the transfer will be downward, and the situation becomes the exact opposite of the one just discussed. On the other hand, if the air is sufficiently dry, a situation arises in which the air gains moisture but, being the warmer body, must surrender heat to the water and thus provide the latent heat of evaporation. The process of evaporation then represents merely a transformation of energy within the atmosphere. However, the evaporation, in accelerating the rate of cooling, will bring the "surface" temperatures of the two media more rapidly into adjustment, so that, over a significant interval, the atmosphere will lose less energy than if no evaporation occurred.

In the diurnal cycle, with the equation of energy balance as in equation (7.4), the contribution going to evaporation from the surface will reduce the total amount to be shared between H_0 and G_0 but should not affect their relative magnitudes. No detailed analytical treatment of the diurnal cycle including the effects of evaporation has been worked out. Accurate assessment of E involves the considerations of chapter 7, but an approximate formula independent of gradients is available, for saturated surfaces only, under certain other ideal assumptions. If the diurnal range is small enough for q_s to be taken as a linear function of T_0, then, to the extent that the mechanisms for vertical diffusion of heat and moisture are the same everywhere, q and $T + \Gamma z$ will be similar functions of height and time, and equation (7.5) becomes

$$\frac{LE}{H_0} = \frac{L}{c_p}\left(\frac{\partial q_s}{\partial T}\right)_{T=T_0}. \tag{8.21}$$

The ratio increases from 0.6 at 270° A to 5.2 at 310° A. In conjunction with equations (7.4) and (8.10), E, H_0, and G_0 are then given separately in terms of R_0 and the bulk properties.

BIBLIOGRAPHY

ANDERSON, E. R., ANDERSON, L. J., and MARCIANO, J. J. 1950. *A Review of Evaporation Theory and Development of Instrumentation.* Report No. 159. San Diego: U.S. Navy Electronics Laboratory.

ANGUS, D. E. 1956. *Measurements of Dew.* Paper No. 36. Australia-UNESCO Symp. Arid Zone Climatology.

BALL, F. K. 1958. "Some Observations of Bent Plumes," *Quart. Jour. Roy. Meteorol. Soc.*, **84**:61–65.

BATCHELOR, G. K. 1953. "The Conditions for Dynamical Similarity of Motions of a Frictionless Perfect-Gas Atmosphere," *Quart. Jour. Roy. Meteorol. Soc.*, **79**:224–35.

———. 1954. "Heat Convection and Buoyancy Effects in Fluids," *ibid.*, **80**:339–58.

BEST, A. C. 1935. *Transfer of Heat and Momentum in the Lowest Layers of the Atmosphere.* Geophysical Mem. No. 65. London: Air Ministry, Meteorol. Office.

BEST, A. C., KNIGHTING, E., PEDLOW, R. H., and STORMONTH, K. 1952. *Temperature and Humidity Gradients in the First 100 M over South-east England.* Geophysical Mem. No. 89. London: Air Ministry, Meteorol. Office.

BOSANQUET, C. H., CAREY, W. F., and HALTON, E. M. 1950. "Dust Deposition from Chimney Stacks," *Proc. Inst. Mech. Eng.*, Nos. 1–4, pp. 355–67.

BUAJITTI, K., and BLACKADAR, A. K. 1957. "Theoretical Studies of Diurnal Wind-Structure Variations in the Planetary Boundary Layer," *Quart. Jour. Roy. Meteorol. Soc.*, **83**:486–500.

BUNKER, A. F. 1955. "Turbulence and Shearing Stresses Measured over the North Atlantic Ocean by an Airplane-Acceleration Technique," *Jour. Meterol.*, **12**:445–55.

———. 1956. "Measurement of Counter-Gradient Heat Flows in the Atmosphere," *Australian Jour. Phys.*, **9**:133–43.

Burke, C. J. 1945. "Transformation of Polar Continental Air to Polar Maritime Air," *Jour. Meteorol.*, **2**:94–112.

Businger, J. A. 1955. "On the Structure of the Atmospheric Surface Layer," *Jour. Meteorol.*, **12**:553–61.

Byers, H. R., and Braham, R. R. 1949. *The Thunderstorm.* U.S. Department of Commerce, Weather Bureau.

Charnock, H., Francis, J. R. D., and Sheppard, P. A. 1956. "An Investigation of Wind Structure in the Trades: Anegada, 1953," *Phil. Trans. Roy. Soc. London, A*, **249**:179–234.

Craddock, J. M. 1951. "The Warming of Arctic Air Masses over the North Atlantic," *Quart. Jour. Roy. Meteorol. Soc.*, **77**:355–64.

Craig, R. A. 1949. "Vertical Eddy Transfer of Heat and Water Vapour in Stable Air," *Jour. Meteorol.*, **6**:123–33.

Cramer, H. E., and Record, F. A. 1953. "The Variation with Height of the Vertical Flux of Heat and Momentum," *Jour. Meteorol.*, **10**:219–26.

Deacon, E. L. 1949. "Vertical Diffusion in the Lowest Layers of the Atmosphere," *Quart. Jour. Roy. Meteorol. Soc.*, **75**:89–103.

———. 1953. *Vertical Profiles of Mean Wind in the Surface Layers of the Atmosphere.* Geophysical Mem., No. 91. London: Air Ministry, Meteorol. Office.

———. 1955. *The Turbulent Transfer of Momentum in the Lowest Layers of the Atmosphere.* Technical Paper No. 4. Melbourne: C.S.I.R.O. Div. Meteorological Physics.

———. 1957*a*. "Wind Profiles and the Shearing Stress—an Anomaly Resolved," *Quart. Jour. Roy. Meteorol. Soc.*, **83**:537–40.

———. 1957*b*. "Eddy Flux Measurements over the Sea," *Proc. Internat. Assoc. Physical Oceanography, Toronto.*

Deacon, E. L., Priestley, C. H. B., and Swinbank, W. C. 1958. *Evaporation and the Water Balance*, pp. 9–34. Washington: UNESCO, Arid Zone Research, No. X: Climatology.

Deacon, E. L., Sheppard, P. A., and Webb, E. K. 1956. "Wind Profiles over the Sea and the Drag at Sea Surface," *Australian Jour. Physics*, **9**:511–41.

Deacon, E. L., and Swinbank, W. C. 1956. "Comparison between Momentum and Water Vapour Transfer." Paper No. 2. Australia-UNESCO Symp. Arid Zone Climatology.

Elliott, W. P. 1957. "A Comparison of Some Approaches to the Diabatic Wind Profile," *Trans. Amer. Geophys. Union*, **38**:21–24.

Ellison, T. H. 1956. "Atmospheric Turbulence," in *Surveys in Mechanics*, pp. 400–430. Cambridge: Cambridge University Press.

Ertel, H. 1942. "The Vertical Turbulent Heat Flux in the Atmosphere," *Meteorol. Zeitschr.*, **59**:250–53.

Ferguson, J. 1952. "The Rate of Natural Evaporation from Shallow Ponds," *Australian Jour. Sci. Res., A*, **5**:315–30.

Flower, W. D. 1937. *An Investigation into the Variation of the Lapse Rate of Temperature in the Atmosphere near the Ground at Ismailia, Egypt.* Geophysical Mem. No. 71. London: Air Ministry Meteorol. Office.

Frankenberger, E. F. H. 1952. "Investigation of the Vertical Exchange within the Lowest Decameters at Quickborn," *Internat. Symp. on Atmospheric Turbulence in the Boundary Layer*, pp. 323–41. U.S.A.F. Geophys. Research Paper No. 19.

———. 1953. "Sommerliche Mittagsprofile der potentiellen Temperatur, der spezifischen Feuchte, der thermischen Energie und der Windstärke an klaren Tagen über einem Schleswig-Holsteiner Wiesengelände bei verschiedenen Windstärken," *Ann. Meteorol.*, **6**:1–4.

Geiger, R. 1948. "Wärmehaushaltskonstanten für den Agrarmeteorologen," *Meteorol. Rundschau*, **1**:322.

Halstead, M. H. 1943. "A Stability-Term in the Wind-Gradient Equation," *Trans. Amer. Geophys. Union*, **24**:204–8.

Hofmann, G. 1953. "Die Thermodynamik der Taubildung," *Ber. Deutsches Wetterdienstes*, Vol. **3**, No. 18.

Holzman, B. 1943. "The Influence of Stability on Evaporation," *Ann. New York Acad. Sci.*, **44**:13–18.

Jacobs, W. C. 1942. "On the Energy Exchange between Sea and Atmosphere," *Jour. Marine Res.*, **5**:37–66.

Kohler, M. A. 1954. "Lake and Pan Evaporation: Water Loss Investigations," *Lake Hefner Studies Technical Report*, **1**:127–48. Geological Survey Prof. Paper No. 269. Washington: U.S. Geological Survey.

Korb, G., Michalowsky, J., and Möller, F. 1956. *Investigations of the Heat Balance of the Troposphere.* Johannes Gutenberg Univ., Meteorol.-Geophys. Inst. Tech. Rept. No. 1.

LAIKHTMAN, D. L. 1944. "Profile of Wind and Interchange in the Layer of the Atmosphere near the Ground," *Bull. Acad. Sci. U.S.S.R.*, Geog. and Geophys. Ser. 8, No. 1, pp. 1–5.

LETTAU, H. 1950. "A Re-examination of the 'Leipzig Wind Profile' Considering Some Relations between Wind and Turbulence in the Frictional Layer," *Tellus*, **2**:125–29.

———. 1951. "Theory of Surface-Temperature and Heat-Transfer Oscillations near a Level Ground Surface," *Trans. Amer. Geophys. Union*, **32**:189–200.

———. 1957. "Windprofil, innere Reibung und Energieumsatz in der unteren 500 M über dem Meer," *Beitr. Phys. d. Atmosphäre*, **30**:78–96.

LETTAU, H., and DAVIDSON, B. 1957. *Exploring the Atmosphere's First Mile*, Vols. **1** and **2**. New York: Pergamon Press.

MACCREADY, P. B. 1953. "Structure of Atmospheric Turbulence," *Jour. Meteorol.*, **10**:434–49.

MCILROY, I. C. 1955. *The Atmospheric Fine Structure Recorder.* Technical Paper No. 3. Melbourne: C.S.I.R.O. Div. Meteorological Physics.

MAL, S., DESAI, B. N., and SIRCAR, S. P. 1942. "An Investigation into the Variation of the Lapse Rate of Temperature in the Atmosphere near the Ground at Drigh Road, Karachi," *Mem. Indian Meteorol. Dept.*, Vol. **29**, Part 1.

MALKUS, J. S. 1957. *On the Structure of the Trade Wind Moist Layer.* Woods Hole Oceanographic Institute, Reference No. 57–59.

MARCIANO, J. J., and HARBECK, G. E. 1952. "Mass Transfer Studies: Water Loss Investigations," *Lake Hefner Studies Technical Report*, **1**:46–70. Geological Survey Prof. Paper No. 269. Washington: U.S. Geological Survey.

MONIN, A. S., and OBUKHOV, A. M. 1954. *Basic Regularity in Turbulent Mixing in the Surface Layer of the Atmosphere.* U.S.S.R. Acad. Sci., Geophys. Inst., No. 24.

MONTEITH, J. L. 1957. "Dew," *Quart. Jour. Roy. Meteorol. Soc.*, **83**: 322–41.

MORTON, B. R. 1957. "Buoyant Plumes in a Moist Atmosphere," *Jour. Fluid Mechanics*, **2**:127–44.

MORTON, B. R., TAYLOR, SIR G. I., and TURNER, J. S. 1956. "Turbulent Gravitational Convection from Maintained and Instantaneous Sources," *Proc. Roy. Soc. London, A*, **234**:1–23.

OBUKHOV, A. M. 1946. "Turbulence in an Atmosphere of Nonhomogeneous Temperature," *Trans. Inst. Theoret. Geophys. U.S.S.R.*, **1**:95–115.

PANOFSKY, H. A. 1953*a*. "The Variation of the Turbulence Spectrum with Height under Superadiabatic Conditions," *Quart. Jour. Roy. Meteorol. Soc.*, **79**:150–53.

———. 1953*b*. *Statistical Properties of the Vertical Flux and Kinetic Energy at 100 Metres*. Pennsylvania State College, Div. of Meteorology, Sci. Rept. No. 2.

PANOFSKY, H. A., and MCCORMICK, R. A. 1954. "Properties of Spectra of Atmospheric Turbulence at 100 Metres," *Quart. Jour. Roy. Meteorol. Soc.*, **80**:546–64.

PASQUILL, F. 1949*a*. "Some Estimates of the Amount and Diurnal Variation of Evaporation from a Clayland Pasture in Fair Spring Weather," *Quart. Jour. Roy. Meteorol. Soc.*, **75**:249–56.

———. 1949*b*. "Eddy Diffusion of Water Vapour and Heat near the Ground," *Proc. Roy. Soc. London, A*, **198**:116–40.

PENMAN, H. L. 1948. "Natural Evaporation from Open Water, Bare Soil and Grass," *Proc. Roy. Soc. London, A*, **193**:120–45.

———. 1956. "Evaporation: An Introductory Survey," *Netherlands Jour. Agr. Sci.*, **4**:9–27.

PEREPELKINA, A. V. 1957. "Some Results of Investigation of the Turbulent Fluctuations of Temperature and of the Vertical Component of Wind Velocity," *Bull. Acad. Sci. U.S.S.R.*, Geophys. Ser. No. 6, pp. 765–78.

PETTERSSEN, S. 1939. "Some Aspects of the Formation and Dissipation of Fog," *Geofys. Pub.*, Vol. **12**, No. 10.

———. 1956. *Weather Analysis and Forecasting*, Vol. **1**. 2d ed. New York: McGraw-Hill Book Co.

POPPENDIEK, H. F. 1952. "A Periodic Heat Transfer Analysis for an Atmosphere in Which the Eddy Diffusivity Varies Sinusoidally with Time and Linearly with Height," *Jour. Meteorol.*, **9**:368–70.

PRANDTL, L. 1932. "Meteorologische Anwendung der Strömungslehre." Beitr. Phys. Atmosphäre, **19**:188–202.

PRIESTLEY, C. H. B. 1953. "Buoyant Motion in a Turbulent Environment," *Australian Jour. Phys.*, **6**:279–90.

———. 1954*a*. "Convection from a Large Horizontal Surface," *ibid.*, **7**:176–201.

———. 1954*b*. Vertical Heat Transfer from Impressed Temperature Fluctuations, *ibid.*, pp. 202–9.

———. 1955. "Free and Forced Convection in the Atmosphere near the Ground," *Quart. Jour. Roy. Meteorol. Soc.*, **81**:139–43.

———. 1956*a*. "A Working Theory of the Bent-Over Plume of Hot Gas," *ibid.*, **82**:165–76.

———. 1956*b*. "Convection from the Earth's Surface," *Proc. Roy. Soc. London, A*, **238**:287–304.

———. 1957. *The Evolution of Energy Gain by the Atmosphere through Contact with the Ground or Ocean*. University of Chicago, Sci. Rept. No. 2. Contract AF 19 (604)-2179. Chicago: University of Chicago Press.

PRIESTLEY, C. H. B., and BALL, F. K. 1955. "Continuous Convection from an Isolated Source of Heat," *Quart. Jour. Roy. Meteorol. Soc.*, **81**:144–57.

PRIESTLEY, C. H. B., and SHEPPARD, P. A. "Turbulence and Transfer Processes in the Atmosphere," *Quart. Jour. Roy. Meteorol. Soc.*, **78**:488–529.

PRIESTLEY, C. H. B., and SWINBANK, W. C. 1947. "Vertical Transport of Heat by Turbulence in the Atmosphere," *Proc. Roy. Soc. London, A*, **189**:543–61.

RAILSTON, W. 1954. "The Temperature Decay Law of a Naturally Convected Air Stream," *Proc. Phys. Soc. London, B*, **67**:42–51.

RICHARDSON, L. F. 1920. "The Supply of Energy from and to Atmospheric Eddies," *Proc. Roy. Soc. London, A*, **97**:354–73.

RIDER, N. E. 1954*a*. "Eddy Diffusion of Momentum, Water Vapour, and Heat near the Ground," *Phil. Trans. Roy. Soc. London, A*, **246**:481–501.

———. 1954*b*. "Evaporation from an Oat Field," *Quart. Jour. Roy. Meteorol. Soc.*, **80**:198–212.

ROBINSON, G. D. 1950. "Two Notes on Temperature Changes in the Troposphere Due to Radiation," *Centenary Proc., Roy. Meteorol. Soc.*, pp. 26–29.

———. 1951. "The Vertical Convective Heat Flux in the Atmosphere—in a Critical Essay," *Quart. Jour. Roy. Meteorol. Soc.*, **77**:61–75.

ROHWER, C. 1931. *Evaporation from Free Water Surfaces*. U.S. Dept. Agriculture, Tech. Bull. 271.

ROSSBY, C. G., and MONTGOMERY, R. B. 1935. *The Layer of Frictional Influence in Wind and Ocean Currents*. Papers in Physical Oceanography and Meteorology, M.I.T. and Woods Hole Oceanographic Institution, Vol. **3**, No. 3.

ROUSE, H., YIH, C. S., and HUMPHREYS, H. W. 1952. "Gravitational Convection from a Boundary Source," *Tellus*, **4**:201–10.

SCHMIDT, F. H. 1957*a*. *On the Diffusion of Stack Gases in the Atmosphere*. Konig. Nederlands Met. Inst., Med. en Verhandelingen. No. 68.

———. 1957*b*. "On the Diffusion of Heated Jets," *Tellus*, **9**:378–83.

SCHMIDT, W. 1918. "Wirkungen des Luftaustausches auf das Klima und den täglichen Gang der Lufttemperatur in der Höhe," *Sitzber. Akad. Wissensch. Wien*, **127**:1942–57.

———. 1941. "Turbulent Expansion of a Stream of Heated Air," *Zeitschr. angew. Math. Mech.*, **21**:265–78 and 351–63.

SHEPPARD, P. A. 1947. "The Aerodynamic Drag of the Earth's Surface and the Value of Von Karman's Constant in the Lower Atmosphere," *Proc. Roy. Soc. London, A*, **188**:208–22.

SHEPPARD, P. A., CHARNOCK, H., and FRANCIS, J. R. D. 1952. "Observations of the Westerlies over the Sea," *Quart. Jour. Roy. Meteorol. Soc.*, **78**:563–82.

SHEPPARD, P. A., and OMAR, M. H. 1952. "The Wind Stress over the Ocean from Observations in the Trades," *Quart. Jour. Roy. Meteorol. Soc.*, **78**:583–89.

STALEY, D. O. 1956. "The Diurnal Temperature Wave for Bounded Eddy Conductivity," *Jour. Meterol.*, **13**:13–20.

SUTTON, O. G. 1934. "Wind Structure and Evaporation in a Turbulent Atmosphere," *Proc. Roy. Soc. London, A*, **146**:701–22.

———. 1948. "Convection in the Atmosphere near the Ground," *Quart. Jour. Roy. Meteorol. Soc.*, **74**:13–30.

———. 1953. *Micrometeorology*. New York: McGraw-Hill Book Co.

SVERDRUP, H. U. 1937. "On the Evaporation from the Oceans," *Jour. Marine Res.*, **1**:3–14.

SVERDRUP, H. U. 1951. "Evaporation from the Oceans," in *Compendium of Meteorology*, pp. 1071–81. Boston: Amer. Met. Soc.

SVERDRUP, H. U., JOHNSON, M. W., and FLEMING, R. H. 1942. *The Oceans*. New York: Prentice Hall, Inc.

SWINBANK, W. C. 1951. "The Measurement of Vertical Transfer of Heat and Water Vapour and Momentum in the Lower Atmosphere with Some Results," *Jour. Meteorol.*, **8**:135–45.

———. 1955. *An Experimental Study of Eddy Transports in the Lower Atmosphere*. Technical Paper No. 2. Melbourne: C.S.I.R.O. Div. Meteorological Physics.

TAIT, G. W. C. 1949. "The Vertical Temperature Gradient in the Lower Atmosphere under Daylight Conditions," *Quart. Jour. Roy. Meteorol. Soc.*, **75**:287–92.

TAYLOR, G. I. 1915. "Eddy Motion in the Atmosphere," *Phil. Trans. Roy. Soc. London, A*, **215**:1–26.

———. 1922. "Diffusion by Continuous Movement," *Proc. London Math. Soc.*, **20**:196–212.

TAYLOR, R. J. 1952. "The Dissipation of Kinetic Energy in the Lowest Layers of the Atmosphere," *Quart. Jour. Roy. Meteorol. Soc.*, **78**:179–85.

———. 1955. "Some Observations of Wind Velocity Autocorrelations in the Lowest Layers of the Atmosphere," *Australian Jour. Phys*, **8**:535–44.

———. 1956*a*. "Some Measurements of Heat Flux at Large Negative Richardson Number," *Quart. Jour. Roy. Meteorol. Soc.*, **82**:89–91.

———. 1956*b*. "An Electronic Analogue of the Saturated Vapour Pressure Curve of Water," *Jour. Scient. Instr.*, **33**:436–38.

———. 1958*a*. "A Linear Unidirectional Anemometer of Rapid Response," *ibid.*, **35**:47–52.

———. 1958*b*. "Thermal Structures in the Lowest Layers of the Atmosphere," *Australian Jour. Phys.*, **11**, 168–76.

TAYLOR, R. J., and DYER, A. J. 1958. "An Instrument for Measuring Evaporation from Natural Surfaces," *Nature*, **181**:408–9.

TAYLOR, R. J., and WEBB, E. K. 1955. *A Mechanical Computer for Micrometeorological Research*. Technical Paper No. 6. Melbourne: C.S.I.R.O. Div. Meteorological Physics.

THOMAS, D. B., and TOWNSEND, A. A. 1957. "Turbulent Convection over a Heated Horizontal Surface," *Jour. Fluid Mechanics*, **2**: 473–92.

THORNTHWAITE, C. W., and HOLZMAN, B. 1939. "The Determination of Evaporation from Land and Water Surfaces," *Monthly Weather Rev.*, **67**:4–11.

THORNTHWAITE, C. W., and KASER, P. 1943. "Wind-Gradient Observations," *Trans. Amer. Geophys. Union*, **24**:166–81.

TOLLMIEN, W. 1926. "Berechnung turbulenter Ausbreitungsvorgange," *Zeitschr. angew. Math. Mech.*, **6**:468–78.

VEHRENCAMP, J. E. 1953. "Experimental Investigation of Heat Transfer at an Air-Earth Surface," *Trans. Amer. Geophys. Union*, **34**:22–30.

VRIES, D. A. DE. 1956. "On the Evolution of Thermal (or Eddy) Diffusivity from the Diurnal Temperature Wave," *Netherlands Jour. Agr. Sci.*, **4**:136–39.

———. 1957. "On the Integration of the Heat-Conduction Equation with Periodic Variation of Temperature," *Jour. Meteorol.*, **14**:71–76.

WEBB, E. K. 1955. *Autocorrelations and Energy Spectra of Atmospheric Turbulence*. Technical Paper No. 5. Melbourne: C.S.I.R.O. Div. Meteorological Physics.

———. 1958. "Vanishing Potential Temperature Gradients in Strong Convection," *Quart. Jour. Roy. Meteorol. Soc.*, **84**:118–25.

WOODCOCK, A. H. 1940. "Convection and Soaring over the Open Sea," *Jour. Marine Res.*, **3**:248–53.

INDEX